# High School and Undergraduate Physics Practicals

# High School and Undergraduate Physics Practicals

## With 3D Simulations

Robert Lucas

### CRC Press

Taylor & Francis Group

Boca Raton  London  New York

CRC Press is an imprint of the
Taylor & Francis Group, an **informa** business

*The front cover illustration shows the simulation of the magnetic flux linkage experiment*

First edition published 2022
by CRC Press
6000 Broken Sound Parkway NW, Suite 300, Boca Raton, FL 33487–2742

and by CRC Press
4 Park Square, Milton Park, Abingdon, Oxon, OX14 4RN

*CRC Press is an imprint of Taylor & Francis Group, LLC*

*Library of Congress Cataloging-in-Publication Data*
A catalog record for this book has been requested

ISBN: 978-1-032-20129-0 (hbk)
ISBN: 978-1-032-19739-5 (pbk)
ISBN: 978-1-003-26235-0 (ebk)

DOI: 10.1201/9781003262350

Access the companion website: https://www.virtual-science.org.uk/

*To Sarah, Henry, and Toby*

# Contents

# About the Author

**Robert Lucas** has worked in and around education for forty years. In addition to teaching at university undergraduate and postgraduate levels, he has created many software products for teaching. These include TOM, the Thoroughly Obedient Moron, a computer simulator for Windows; the Virtual Language Laboratory, which was the first multimedia language lab that used a PC network; and many science 3D simulations based on his research into online methods for teaching physics.

# Introduction

## Background

The importance of performing practicals or experiments in Physics is fundamental to an appreciation of science and its history. It shows us how we arrive at Physical laws from the collection of data that correspond to our observations. It underpins our notions of discovery and imposes a certain discipline on our endeavors to understand the world we live in by showing us how we can test a hypothesis in a rigorous manner using what is often described as a controlled experiment, where we are careful to eliminate all unwanted influences on the experimental data that are collected. The research philosophy that this involves is called logical positivism, which is entirely based on what can be observed and measured, and then using this to verify or nullify hypotheses which often are generalizations, taking the form of a Physical law such as Newton's law of gravitation. We don't necessarily expect our laws to be absolute, but they may become refined or even replaced as we make better and more accurate measurements. This has happened with Newton's law of gravity, where a measured anomaly in Mercury's orbit couldn't be explained until Einstein's theory of general relativity. However, due to the simplicity and the accuracy of Newton's law in most cases, we still use it while being fully aware of its limitations.

## This Book

Each chapter in the book describes an experiment. In general, there is background information, a description of the apparatus, a statement of the aim, what variables are involved, the method for doing the experiment, some notes on the expected results, and some further discussion. Each practical described in this book can be performed within an actual laboratory with real equipment, or it can be performed as a 3D simulation on a computer. In either case, the experiment should be reported in detail in the student's laboratory book. It is assumed that with the real experiments, these are performed with adequate supervision and any safety concerns are dealt with comprehensively. Clearly there are no safety concerns with any of the simulations, which are nevertheless highly realistic and don't just superficially

DOI: 10.1201/9781003262350-1

get good results, as with a real experiment. For example, a magnetic needle will take time to settle down, and you will have to put your 'eye' in a good position to avoid parallax errors. Where possible, the apparatus mimics a piece of actual hardware; for example, the power supply commonly used is modeled on a common Maplins model, and the oscilloscope used is one commonly used in education.

When you have completed the experiments herein, you should be in a strong position to plan and execute experiments of your own and take charge of making your own discoveries. The methods learned here will, of course, help you get good grades on your Physics exams, but they are of much wider application and should be considered as an extremely valuable life skill. For example, the scientific mindset gained might help you diagnose a fault in any mechanical or electromechanical device, which could be your car or motorcycle or a simple everyday appliance, by proposing and testing hypotheses.

The individual chapters do not describe the recording of inherent uncertainties in measurement, nor how these might be processed, as this would add unnecessarily to the text; however, there is an appendix on recording and processing your data while taking account of any uncertainty in measurements.

## The Scientific Approach

The very first thing we need to do is give our investigation a title so that we can unambiguously refer to it. Then we must state precisely:

1. What it is we are trying to establish.
2. What variables are involved.
3. Which variable we are intending to change.
4. Which variables we are intending to keep the same.
5. What method we are going to use.
6. What equipment is needed.
7. How many observations or measurements we intend to make.

In most cases, when we have collected our data set, we plot it and then discuss what conclusions can be drawn from it, if any, by careful analysis of the results. This may involve calculating a best-fit line.

The very act of stating what we are trying to do can be extremely useful in creating a clear picture of what we are attempting. What it is we are trying to establish is often written as a hypothesis or a question that we can test the answer to. So a hypothesis might be 'The pressure of a gas is proportional to the temperature when the volume is constant'. This could also be couched as

'Find the relationship between the pressure and the temperature of a gas of constant volume'.

Identifying the variables that are involved is very much part of the investigative skills necessary when conducting experiments. Being careful about the variables that are involved is often critical, as there can be many variables that influence a result and it is essential to only be varying one at a time, otherwise our results will be worthless. It can be quite easy to overlook the significance of a variable that might not be under our direct control, such as background light, noise, or room temperature interfering with our observations. We need to be aware of all the variables that can affect our results so that if we cannot find a way of factoring their influence out of the experiment, we can at least discuss what effect the interference may have had and how the conclusions might be interpreted in the light of any such effects.

Being clear, unambiguous, rigorous, and precise adds value to our results that can be appreciated by others. In the scientific world, others will want to validate our results, and for this they need a clear description of the methodology so that if different results are obtained, these can be effectively investigated.

## Recording and Processing Uncertainties

Every measurement will have an inherent uncertainty that affects the accuracy of any result obtained. The use of uncertainties is not present in any of the experiments' descriptions in this book, as this would bloat each chapter unnecessarily. However, Appendix 1 shows how to record, process, and use the uncertainties within your results. Although this may seem like an additional burden, it is very easily handled if you collect and process your data in a spreadsheet.

## Using a Spreadsheet

Many of these experiments are aided enormously by recording the observations in a spreadsheet and taking advantage of being able to repeat calculations for many rows of data. Given the convenience of this, I have included Appendix 2 on how to use Microsoft Excel to record and analyze results.

## Controlling the Simulations

Although all of the chapters contain notes on using the simulations, which can be controlled either with a mouse or by using on-screen controls, each method is not always described within each chapter. So these two ways of controlling the simulations are dealt with in detail in Appendix 3.

## Thanks

Special thanks must go to Chris Dimmer for ongoing support throughout this project, and to Martin Spann for testing the original set of experiments on the Mac.

**Rob Lucas**
*Chippenham*
*September 2021*

# 1

## Millikan's Oil Drop Experiment

## Introduction

This experiment is one of the most important ever performed. Joseph John 'J. J.' Thomson had identified the electron as a charged particle in 1897 and had measured its charge-to-mass ratio. The next step was to obtain a value for either the mass or the charge.

Robert Millikan was born in Illinois in 1868 and obtained his Physics doctorate in 1895, just two years before Thomson's experiment. He received the Nobel Prize in 1923 'for his work on the elementary charge of electricity and on the photoelectric effect'.

In Millikan's experiment, he sprayed oil drops into a chamber (C). These drops obtain a charge from the friction of going through the atomizer at (A). These drops fall under gravity through a hole in the plate (M). There is another plate (N) below the first plate, and these are separated by a ring with several windows. One of these windows is used to illuminate the chamber between the two plates, and another window is used to observe the motion of the oil drops.

An electrostatic charge can be applied to the plates. This charge can be varied, and it allowed Millikan to balance the force due to gravity against the electrostatic force and bring the particles to a standstill. It is this balancing of forces that allowed Millikan to determine the charge on an electron. One small detail remains, and that is that the oil drops had a number of charges on them that only sometimes would be due to a single electron. So Millikan observed the motion of particles with multiples of the electron charge. This meant that he had to do many observations and look for the ones that had the least charge on to find the single charge amount.

In our version of the experiment, tiny (1.02 microns across) polymer balls are used instead of oil. We use an equation of motion that balances the forces due to gravity, drag, and the electrostatic. As with Millikan's version, we get a distribution of the number of charged electrons on our polystyrene balls.

- **Question**: What is the volume of a single polymer ball?
- **Question**: What is the mass of a single polymer ball given that the density is 1.05 g/ml?

DOI: 10.1201/9781003262350-2

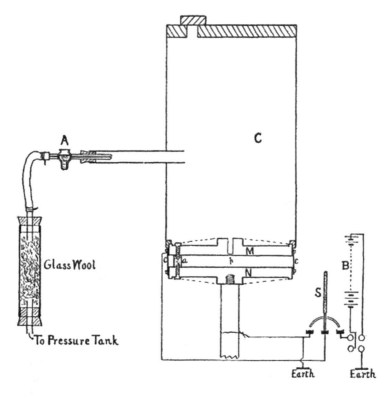

**FIGURE 1.1**
Millikan's drawing of his apparatus.

## The Objective

To find the charge on a single electron.

## The Apparatus

The apparatus is shown in Figure 1.2.
    You will need:

- A high-voltage power supply for the plates with polarity and voltage
  control
- A low-voltage power supply for the light bulb
- An electrostatic chamber
- A multimeter
- A bottle of polymer balls

- A squeeze atomizer
- A light source
- A telescope or webcam with reticle
- A micrometer

The kit can either be sourced in the form shown in Figure 1.2 or assembled from the components listed. The electrostatic chamber is the small box behind the webcam and to the left of the light. In this case, it was supplied as part of the main unit, but it can be sourced as a separate item. It consists of two metal plates held apart by a transparent separation ring. The metal plates are fitted with connectors so that they can be connected to a high-voltage electrical supply. There are three other holes into the internal chamber: one for introducing the particles, one for letting light in so that the particles are illuminated, and one for observing the interior of the chamber.

**FIGURE 1.2**
View of the experiment's simulation.

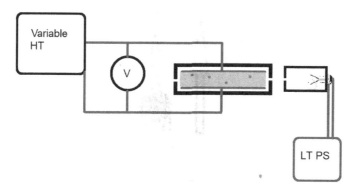

**FIGURE 1.3**
Schematic of the electrical circuits.

The space between the plates is viewed using a webcam, although it can also be viewed directly by eye through a telescope. This is considerably less convenient due to how easy it is to disturb the equipment. The image is displayed on the screen when the chamber is illuminated, and the webcam is activated. The high- and low-voltage supply is the box behind the chamber. To the right of the chamber, you can see the light source. The low-voltage output is connected to the light. The high-voltage output goes to the box that controls the polarity and voltage. A meter to the left of the power supply indicates the voltage difference between the plates. There are controls on the unit below the chamber for varying the voltage and reversing its polarity.

The atomizer is to the left of the electrostatic chamber and is filled with a solution containing the polymer balls. Squeezing the bulb of the atomizer sends a puff of balls into the chamber.

## The Variables

The independent variable is the voltage, and we are going to measure the speed of the polymer particles, which is the dependent variable.

## The Physics

When the polymer particles are moving inside the chamber, they can be considered to be at their terminal velocity. Although for some of the time they are accelerating because they are so light, they reach their terminal velocity almost immediately. Therefore, in all cases we can consider that the forces acting on the particles cancel out; that is, upward forces equal downward forces. We are going to observe the velocities of the polymer balls as they move upwards. In this case:

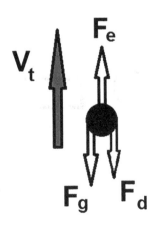

**FIGURE 1.4**
The forces acting on a polymer sphere.

$V_t$ is the terminal velocity, which in this case (and the cases we are going to observe) is an upward drift. The upward force on the particle is $F_e$, the electrostatic force. The downward forces are $F_g$, the force due to gravity, and $F_d$, the force due to viscous drag. The forces balance, so we can write:

$$F_e = F_g + F_d$$

The electrostatic force is given by:

$$F_e = qE$$

Where E is the field, which in this case is simply described as the voltage divided by the distance between the plates. Therefore:

$$F_e = qV/d$$

Where d is the distance between the plates. The drag on a sphere is given by Stokes' Law, which states:

$$F_v = 6\pi r\eta V_t$$

Where $r$ is the radius of the sphere and $\eta$ is the viscosity of air ($1.81 \times 10^{-5}$ N sec/m).
   The force due to gravity is given by:

$$F_g = mg$$

Where $m$ is the mass of the polymer ball and $g$ is the acceleration due to gravity (9.81 m/s²).

- **Question**: What is the force on one of the polymer balls due to gravity in Newtons?

---

## The Method

The first thing you need to do is find the separation of the plates in the electrostatic chamber. You may simply be given this value, or you may be able to remove the separating ring for measurement with a micrometer.
   Whether you are using a real setup or the simulator, be sure to note your reading as the plate separation, d, in your laboratory book.
   When you have finished measuring the height of the ring, reassemble the chamber.
   Turn on the power supplies to the light and to the plates in the chamber.
   You will see the light go on and an image will appear on the computer monitor. If you are using a telescope, you should be able to see inside the chamber a similar view to that shown here on the monitor.

Using the voltage control, turn the polarity to positive and the voltage to about 300 volts; you may need to use a different voltage for your particular chamber, so refer to the manufacturer's instructions.

Now you can introduce some polymer spheres into the chamber by squeezing on the bulb of the atomizer.

You will now see a collection of white spots on the screen or through the telescope. These are the magnified images of the polymer spheres. You can change polarity and see them fall rapidly (at various speeds, depending on how much charge they have), or you can switch off the voltage and see the particles fall at the same rate due to gravity.

With the voltage set at around 300 volts (you may have a different ideal voltage for your chamber), time the polymer spheres across the divisions seen in the reticule. You need to know the distance that each division in the reticle represents, so that these timings will allow you to calculate the polymer spheres' terminal velocity.

## The Simulation

With the simulation, you can 'explode' the chamber into its constituent parts by clicking on it. When you do this, a micrometer will automatically move to measure the ring that separates the plates. The chamber will only 'explode' when the power is switched off. You will see this:

**FIGURE 1.5**
An exploded view of the electrostatic chamber.

**FIGURE 1.6**
Measuring the thickness of the separating ring with a micrometer.

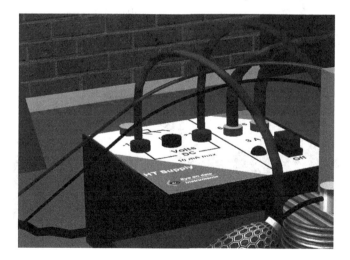

**FIGURE 1.7**
The power supply for high and low tensions.

From the top, you can see the retaining bolts, the top plate, the separating ring with the micrometer, and the bottom plate, which is just above the body of the chamber. You must now rotate the barrel of the gauge to obtain an accurate reading for the height of the ring, which corresponds to the separation of the plates.

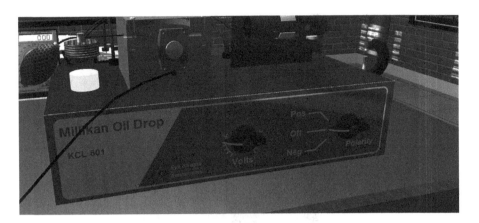

**FIGURE 1.8**
The main control panel.

**FIGURE 1.9**
The squeeze atomizer for introducing the polymer spheres into the chamber.

You can zoom in and out using the plus and minus keys (or Z/X) on the number pad or main keyboard. With the mouse over the rotating barrel of the micrometer, use the mouse wheel to tighten the jaws of the micrometer onto the separating ring. When the barrel will not move any farther, you can then read the micrometer.

When you have finished measuring the width with the micrometer, a second click on the chamber will remove the micrometer and reassemble the chamber, which will enable you to apply a voltage to the plates.

You can switch on the power supply by clicking on its on/off button at the lower right of the unit.

You can change the voltage and polarity using the switches on the main unit beneath the chamber.

You can read the voltage from the meter next to the power supply.

Clicking on the bulb of the atomizer with the left mouse button will introduce the polymer spheres into the chamber.

In the simulation, each division in the reticle is $2.5 \times 10^{-4}$ m.

## The Results

Create a table with headings for the time, the number of divisions, the velocity in divisions per second, the velocity in meters per second, the force due to drag, and the charge. Your table should look like this:

| Time/s | Divisions | v div/s | v/ ms$^{-1}$ | F drag/N | q/coulombs |
|--------|-----------|---------|-----------|----------|------------|
| 15.14  | 12        | 0.792602 | 0.000198 |          |            |
| 14.86  | 10        | 0.672948 | 0.000168 |          |            |
| 72.5   | 10        | 0.137931 | 3.45E-05 |          |            |

You can calculate the drag from this formula:

$$F_v = 6\pi r \eta V_t$$

Notice that this depends on the terminal velocity, which you have calculated in the preceding column. You know the force due to gravity; you should have calculated this in a previous section ($F_g = mg$).

Rearrange:

$$F_e = qV/d$$

To give a formula for q:

$$q = d/VF_e$$

Now substitute $F_e$ from this formula:

$$F_e = F_g + F_d$$

To give a formula for the charge q:

$$q = d/(F_g + F_d)$$

This is the formula to use in the final column of the table.

Time forty or fifty different polymer spheres and you will get a distribution of charges, which are all multiples of the electron charge. A careful examination of this distribution should reveal the actual charge of a single electron.

## Further Discussion

Electrons should repel each other, so why do you think the electrons tend to clump together in this experiment?

This experiment used a webcam to make it easier to view the inside of the chamber without jogging it. Can you think of any other experiments where the use of a webcam might be a good idea for different reasons?

# 2

## Planck's Constant

### Introduction

Planck's constant is a universal constant that has its roots in Max Planck realizing that energy levels could not be continuous but had to be discrete in units of this constant. He had been trying to find a radiation law for black body radiation. Lord Raleigh and James Jeans had proposed a scheme that had worked with the kinetic theory of gases, but when waves were used instead of gas particles, it was found that it predicted an infinite amount of energy at high frequencies. This was named the 'ultraviolet catastrophe', and it is exactly what Planck was trying to eliminate by proposing a new radiation law. The old Raleigh and Jeans way of solving it was to assume the equipartition (equal sharing) of energy between all possible frequencies. However, with waves, you can just keep on fitting more and more into the black body at higher and higher frequencies.

Planck had a great deal of confidence in the experimental results concerning black body radiation, so he became determined to find a law that agreed with

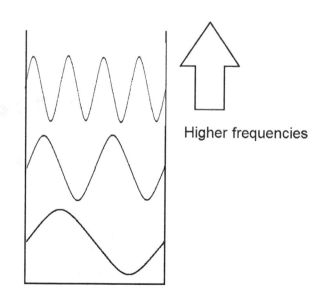

Higher frequencies

**FIGURE 2.1**
A series of higher-frequency waves inside a black body.

DOI: 10.1201/9781003262350-3

the available observations. He eventually found one but had little idea of why it worked. In his quest to justify it, he reworked some of Ludwig Boltzmann's ideas on entropy from his kinetic theory of gases. It was Boltzmann's idea of dividing the energy into small but finite levels that was key. Planck called the amounts of energy that could be radiated 'quanta', and these were given by:

$$E = hf$$

Where $E$ is the energy quantum, $h$ is Planck's constant, and $f$ is the frequency of the radiation. This equation is one of the cornerstones of what was (very rapidly) to become the field of quantum mechanics.

It is possible to measure Planck's constant using some relatively simple equipment. A light-emitting diode (LED) will start to emit light of a particular wavelength at a precise voltage, called the threshold voltage, which can easily be measured. The voltage allows us to calculate the energy going into the LED, and the wavelength of the light allows us to calculate the light's energy. Given that the energy of light is Planck's constant times the frequency (see the earlier equation), we can calculate the value for Planck's constant. There is just one slight complication: energy from heat is also going into the LED. We can eliminate this effect by taking the threshold voltage for different-colored LEDs and assuming that the thermal energy supplied to each LED is the same.

## The Objective

To obtain a value for Planck's constant.

## The Apparatus

The circuit is constructed on a breadboard, as shown in Figure 2.2.

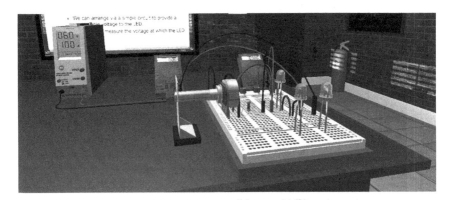

**FIGURE 2.2**
The apparatus showing the power supply, the amp and volt meters, and the breadboard containing all the other components.

You will need:

- A power supply
- Two multimeters, or one ammeter and one voltmeter
- Three different-colored LEDs
- A breadboard
- One 4.7k ohm variable resistor
- One 330 ohm resistor
- Some connecting wire

A variable resistor (also called a potentiometer) on the breadboard controls the voltage going across the LED; this is measured by the multimeter (the device to the right of the power supply). A second multimeter, to the right of the first, is set to measure up to 2 ma (2000 μa).

## The Circuit

The circuit implemented on the breadboard simply allows us to vary and measure the voltage being supplied to the LED and the current flowing through it. The circuit is shown in Figure 2.3.

This shows a 6-volt drop from the power supply across the variable resistor. A lower voltage can then be tapped off this and is provided across the LED in series with a 330 ohm resistor. This resistor limits the current flowing through the LED, which would burn it out. The 'V' in the circle represents our multimeter set to a voltage scale and enables us to measure the voltage across the LED. The 'A' in the circle represents our ammeter.

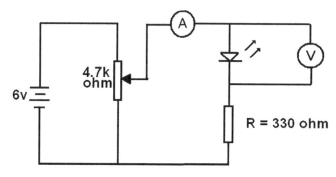

**FIGURE 2.3**
Schematic for the electrical components of the circuit used in the experiment.

## The Variables

The voltage is the independent variable. The current is a dependent variable. A variable that might affect the experiment is the temperature of the room, which will be heating the LED.

## The Physics

The energy of one photon is given by:

$$E_{photon} = hf$$

Where $h$ is Planck's constant and $f$ is the frequency of the photon.
    An electron has energy:

$$E_{electron} = eV$$

Where $e$ is the charge on a single electron ($1.602 \times 10^{-19}$ C) and $V$ is the voltage.
    The energy of the photon can be equated to the energy given to the electron by the voltage and by the thermal energy. There is also going to be some energy lost in the circuit. Hence:

$$hf + \Phi_{loss} = eV + \Phi_{thermal}$$

This can be rearranged to give:

$$eV = hf + \Phi_{loss} - \Phi_{thermal}$$

Frequency multiplied by wavelength is equal to the speed of light, or:

$$f\lambda = c$$

Where $\lambda$ is the wavelength and $c$ is the speed of light. Therefore, the frequency can be expressed in terms of the wavelength as:

$$f = c/\lambda$$

Substituting this into our previous equation for Planck's constant gives us:

$$eV = h\,c/\lambda + \Phi_{loss} - \Phi_{thermal}$$

Finally, we divide by the charge on the electron $e$ to get:

$$V = h\, c/e\lambda + (\Phi_{loss} - \Phi_{thermal})/e$$

The second term on the right can be treated as a single constant, so we can write:

$$V = h\, c/e\lambda + k$$

This is the equation that we will use to calculate a value for Planck's constant. The form of this equation is that of a straight line (think of $y = mx + c$) with $V$ dependent on $1/\lambda$ with a gradient of $hc/e$ and an intercept $k$.

Therefore, if we plot $V$ against $1/\lambda$, the gradient will give us a value for $hc/e$ from which we can easily calculate $h$. Notice that this eliminates the need to know the thermal gains and the electrical losses in the circuit.

---

## The Method

Set the voltage to 6V with a maximum current set to 1A. Switch on the power to the circuit by clicking on the power button on the power supply. The multimeter measuring the voltage should be on a setting for measuring up to 10V, and the second multimeter should be on a setting to measure a few milliamps.

If you want to see the LED light up as soon as possible, then switch the room lights off. In the real world, you might use a dark blanket over the equipment to cut out extraneous light. This will make viewing the lighting of the LED easier. However, we are going to determine the point at which the LED starts to emit photons by monitoring the current, as it is not possible to see the light from this immediately that the LED starts emitting; the number of photons is just too small. What we observe is that the current being measured by our ammeter (the multimeter on the right) will go up sharply at the point when the LED starts to emit photons.

Position yourself so that you can see the multimeters, the LED, and the variable resistor, so your view might be something like that shown in Figure 2.4.

Turn the potentiometer until the reading on the right-hand multimeter goes up sharply. Move the potentiometer forwards and backwards a little to identify the point at which the current starts as accurately as possible. If you increase the voltage a little further, you should see the light from the LED.

Note the voltage indicated on the left-hand multimeter. This is the threshold voltage.

Repeat the experiment for both the green and blue LEDs, noting the threshold voltages and the wavelengths.

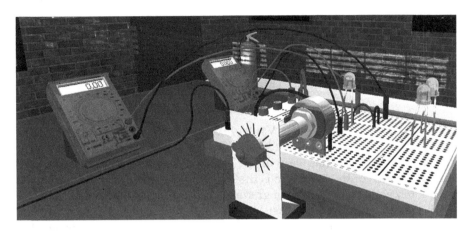

**FIGURE 2.4**
A view of the experiment which allows the reading of necessary values.

## The Simulation

In the simulator, you can switch on the power supply by clicking the on/off button. You should see that the circuit is on by the indicator on the power supply. This is all you need to do with the power supply, as the voltage of 6V has been preset.

The room lights can be switched off by clicking on the switch by the Exit.

The variable resistor can be varied by placing the mouse cursor over the knob and using the mouse wheel or P/N.

There are three LEDs available: red, green, and blue. The red one starts off as part of the circuit. The blue and green ones are simply parked in an unused area of the breadboard. You can change the LED being used in the circuit by clicking on one of the others. For example, if you click on the green LED, then the red LED will be parked in an unused area of the breadboard and the green LED will be positioned in the circuit where the red one was previously positioned. Figure 2.6 shows that the red LED has been parked and the green LED is on the way to its position in the circuit.

Before you can swap an LED, you must switch off the power.

## The Results

Create a table in your laboratory book or in a spreadsheet program with a column for the color of the LED, a column for the wavelength, and a column for the threshold voltage.

Plot a graph of $V$ against the inverse of the wavelength and measure the gradient. Equate this with $hc/e$ and calculate a value for $h$.

**FIGURE 2.5**
The power supply showing the circuit completed setting.

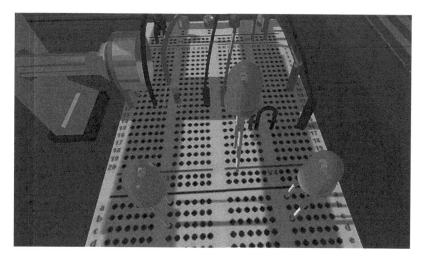

**FIGURE 2.6**
The green LED being moved to the circuit while the red LED is parked.

## Further Discussion

Can you find other methods that can be used to determine Planck's constant? Are they more accurate or less accurate? Why?

# 3

## Rutherford's Gold Foil Experiment

### Introduction

This experiment is one of the most famous, as it revolutionized our ideas of the atom. Before Hans Geiger and Ernest Marsden performed this experiment under the direction of Ernest Rutherford, the atom was known to be made of positively and negatively charged particles but its structure was not known. It was generally accepted to be like J. J. Thomson's suggestion of a plum pudding, with the positive and negative charges – like the fruit in the pudding – being interspersed throughout. Thomson had discovered the electron in 1897.

This is known as Thomson's plum pudding model, and there was some logic behind it. The charges of the same sign could hardly clump together because of the very strong electrical force that repels charges of the same sign.

Rutherford's experiment (1911) was very simple in its construction, if rather harder to perform. It consisted of a source of alpha particles aimed at a thin gold foil and a screen made of zinc sulfide, which scintillated when a particle hit it; that is, it created a small flash of light. The experimenters used a microscope with a screen attached as a viewer.

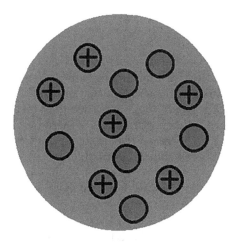

**FIGURE 3.1**
Thomson's plum pudding model of the atom.

DOI: 10.1201/9781003262350-4

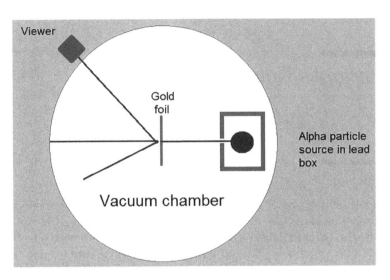

**FIGURE 3.2**
Schematic of Rutherford's experiment.

- Question: What is an alpha particle?

An alpha particle source in a lead box with a small hole creates a narrow beam of particles that are aimed at the thin sheet of gold foil in the canter of the apparatus. The apparatus (except for the detector) is within an evacuated chamber to prevent the alpha particles hitting molecules in the air.

- Question: What is scintillation?

The viewer, consisting of the microscope and zinc sulfide screen, can be placed anywhere in a circle around the apparatus to view the scintillations. The experimenters performed their observations in the dark after a considerable time taken for their eyes to adjust. They then had to count the scintillations, which brought its own difficulties – there is an upper limit to just how fast you can count. Therefore, the apparatus had to be reconfigured for different angles to make the counting possible.

- Question: How might the experiment be configured to slow the speed of the necessary counting?

The experimenters found that although the vast majority of particles went straight through the gold foil without changing direction, a significant number of them were deflected at angles larger than 90 degrees, which simply could not be explained by the plum pudding model. It is relatively easy to calculate the deflection that such a model might produce, and the probability of a deflection of 90 degrees or more is 'vanishingly small'. (See Geiger and Marsden in *The Laws of Deflexion of α Particles Through Large Angles*. These are the two who performed the experiment; it was Rutherford who interpreted the results. You can find this paper online and there is a mistake in the table given on page 7; see if you can spot it.) Rutherford put it rather colorfully as 'It was as if you fired a 15 inch shell at a sheet of tissue paper and it came back to hit you'.[1]

Rutherford deduced from the experiment that the atom had a positively charged center that contained most of the mass of the atom; this he called the nucleus. Alpha particles are positively charged and so must be repelled by a concentration of positive charges, i.e., the proposed nucleus. As relatively few alpha particles were deflected by the gold foil and very few indeed were bounced back, Rutherford deduced that the nucleus must be very small indeed.

## The Objective

To verify that the particle counts at different angles conform to the Rutherford model of the atom.

## The Apparatus

You will need:

- A large vacuum chamber
- A vacuum pump and connecting pipe
- A radioactive source of Americium in a brass container with aperture
- A gold foil
- Moveable particle detector
- Electronic counter
- Scale in degrees

Teachers, technicians, and students must be aware of all regulations involved in handling radioactive substances. Naturally, there are no safety concerns with using the simulation.

The assembled apparatus can be seen in Figure 3.3. Notice the mechanism that is essential for moving the detector around the base of the vacuum chamber to different angles.

**FIGURE 3.3**
View of Rutherford's experiment.

## The Variables

The independent variable is the angle of the detector. The dependent variable is the rate of particles reaching the detector.

## The Physics

Rutherford proposed that the number of particles falling onto a unit area of a screen per second (call this flux) at an angle of $\varphi$ is given by this formula:

$$\text{Flux} = Qntb^4 \, \text{cosec}^4(\varphi/2)/16r^2$$

Question: Find another way of expressing *cosec()* in terms of trigonometric functions that you are familiar with.

Where $Q$ is the number of alpha particles per second emitted, $n$ is the number of atoms per unit volume in the foil, $r$ is the distance from the foil to the detector, and $t$ is the thickness of the foil. The quantity $b$ is given by:

$$b = 2NeE/mu^2$$

*Ne* is the central charge of the atom, *m* is the mass of the alpha particle, and *E* is its charge and *u* its velocity.

We need not concern ourselves too much with the detail. The part we are going to concentrate on is that the count of particles in a given time, call this *N*, is proportional to the fourth power of the cosec of half the angle. That is:

$$N = k \, \text{cosec}^4(\varphi/2)$$

Where *k* is a constant. You will establish that this relationship holds.

## The Method

Start the pump and wait several minutes for the evacuation to complete.

For angles of 1 degree to 10 degrees, count the number of hits on the detector. You need to reset the detector and stop the counting between each set.

## The Simulator

In the simulator, the vacuum pump on the right runs the entire time that you are running the experiment, so no further thought needs to be given to it.

You can start, stop, and reset the counter using the controls on the counter. Place the mouse over the button (which will change color to yellow) and click with the left mouse button.

The red wheel under the table supporting the evacuated dome turns the detector. To turn it, place the mouse over the red wheel (the color will change to yellow) and use the mouse wheel.

The base of the dome has a scale from which the angle can be read. You may need to zoom in/out using the +/- keys on the number pad or Z/X to get a clear reading.

**FIGURE 3.4**
View of the counter showing its controls.

**FIGURE 3.5**
View of the angle scale of the detector.

## The Results

Create a table that has a column for the angle, a column for the count, a column for half the angle, a column for this angle in radians (if your sin calculation is going to need radians), a column for sine of half the angle, a column for cosec of half the angle, and finally, a column for the cosec raised to the power of 4. You may find it convenient to use a spreadsheet program.

| Phi | N/Count min⁻¹ | 0.5phi/deg | 0.5phi/rad | sin(phi/2) | cosec(phi/2) | (cosec(phi/2))^4 |
|-----|---------------|------------|------------|------------|--------------|------------------|
| 1   | 6454480       | 0.5        | 0.008726646 | 0.008727  | 114.593      | 172437834.9      |
| 2   |               |            |            |            |              |                  |
|     |               |            |            |            |              |                  |

Create a row for the angle 1 through 10. Using the apparatus, fill in the count values for each angle. Then fill in the remaining columns for each row.

To show the relationship, we need to plot the second column against the last column and verify that this is a straight line. Unfortunately, because the range of data is so large, this is hard to do. Create a new table that has columns for the logs of the two rows, like this:

| log(N)       | log csc^4 |
|--------------|-----------|
| 6.809861259  | 8.236633  |
|              |           |
|              |           |
|              |           |

Now we can plot these and verify that what we get is a straight line, which shows that Rutherford's relationship between the count and the fourth power of the cosec of half the angle is correct.

## Further Discussion

Rutherford was famously reported to have said 'All science is either Physics or stamp collecting'.[2] What do you think about this comment? To what extent do you think that it is true? What does it say about Rutherford? What is he trying to say about Physics?

## Notes

1. Andrade, D. E. N. C. Rutherford and the Nature of the Atom (Science study series), (1st ed.) (Doubleday, 1964).
2. Bernal, J. *The Social Function of Science* (London: The M.I.T. Press, 1973).

# 4

## Measuring the Acceleration Due to Gravity

### Introduction

Isaac Newton was the first person to realize that one force unites the movements of objects on Earth and out in space. His law of gravitation states that all objects in the universe exert an attractive force on all other objects in the universe, and that this force is proportional to the mass of each object and is inversely proportional to the square of the distance between them. That is:

$$F = GMm/r^2$$

$F$ is the force, $G$ is the gravitational constant, $M$ and $m$ are the masses of the objects, and $r$ is the distance between them. The starting point for the discovery of this law was in divining how the planets moved. Tycho Brahe laid the foundation for this discovery by making accurate observations of the planets' motions, which enabled Johannes Kepler to discover his three Laws of Planetary Motion:

1. The planets move in elliptical orbits with the Sun at one focus.
2. The line joining the planet to the Sun sweeps out equal areas in equal times.
3. The square of the orbital period, T, is proportional to the cube of the mean distance from the Sun, or $T^2 \sim r^3$.

A conker (horse chestnut) being swung on a string does not fly off at a tangent because a force towards the center is being supplied by the swinger. The conker is constantly accelerating towards the center of the circle. Its direction is constantly being bent towards the middle of the circle by the tension in the string.

Around the same time as Kepler was formulating his laws, Galileo was making progress on the movement of bodies. He discovered the principle of inertia, which is if a body is moving in a straight line and has no forces acting on it, it just carries on moving in the same straight line at a uniform speed. Newton added to this, saying that if a body is to speed up (accelerate),

DOI: 10.1201/9781003262350-5

then a force must be applied in the direction of motion and that the heavier the body, the bigger the force must be to get the same effect. The equation (Newton's second law) is:

$$\text{Force} = \text{Mass} \times \text{Acceleration.}$$
$$\text{Or } F = Ma.$$

So when Newton saw an apple fall downwards, he knew that there had to be a force causing this motion, and he wondered if that same force was acting on the Moon. This was an extraordinary, insightful thought.

Eureka! But Newton still needed a formula for calculating the force due to gravity. He realized that a consequence of Kepler's third law is that the force must be weaker farther away from the Sun and directly in proportion to the square of the distance.

Additionally, it seemed clear that heavier objects exerted larger forces, and each body acted on the other. Therefore, the force should be proportional to the product of the two masses. Hence:

$$F = GMm/r^2$$

Where $G$ is the gravitational constant, $M$ and $m$ are the two masses, and $r$ is the distance.

Thus, the Universal Theory of Gravitation was born, which on Earth causes all bodies to accelerate towards the center of the Earth until something gets in the way. The rate of this acceleration is denoted by $g$ and is the subject of this experiment.

Newton was never entirely happy that there was no obvious mechanism by which gravity worked. He wrote:

> That one body may act upon another at a distance through a vacuum, without the mediation of anything else, by and through which their action and force may be conveyed from one to another, is to me so great an absurdity, that I believe no man, who has in philosophical matters a competent faculty of thinking, can ever fall into it.[1]

However, Newton's theory has stood the test of time. Einstein's General Theory of Relativity has very successfully replaced the action at a distance problem with a geometrical solution – matter bends space, and objects follow straight paths in this curved space. However, it is still not clear whether Newton's philosophical objections have quite gone away. How is space told how to bend, and how does this propagate? There are also several gravitational anomalies that have yet to be explained.

## The Objective

To find out the acceleration on Earth due to gravity. In the simulator, we can actually repeat the experiment on the Moon and on Mars.

## The Apparatus

You will need:

- An electromagnet
- A gantry with a trap door switch
- An electronic timer
- A power supply
- A two-pole switch

The apparatus must be wired so that the two-pole switch simultaneously starts the timer and breaks the circuit to the electromagnet.

## The Variables

There is one dependent variable in this experiment: the time.

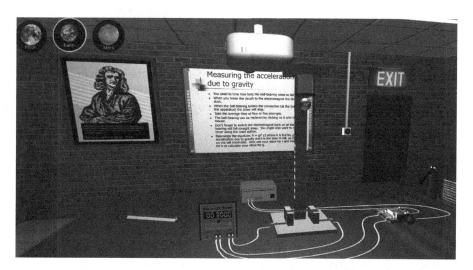

**FIGURE 4.1**
The apparatus for measuring acceleration due to gravity.

## The Physics

For an accelerating body, the distance traveled is given by the formula:

$$s = ut + ft^2/2$$

Where $u$ is the initial speed, which in this case is zero (the ball bearing starts at rest, being held by the electromagnet). $f$ is the acceleration, which in this case is the unknown called $g$, and $t$ is the time. We can rearrange this expression to give us:

$$g = 2s/t^2$$

We can measure the height fallen, that is a given, and $t$ is the time of descent as recorded on the timer.

## The Method

With the switch in the closed position, electricity will flow from the power supply to the electromagnet at the top of the apparatus and keep the ball bearing at this position.

Measure and take note of the distance of the ball bearing to the trap beneath it.

Open the two-pole switch; the ball bearing will drop and the timer will start.

When the ball bearing falls through the trap at the bottom of the apparatus, it breaks the timer circuit and the timer stops. This will now indicate the time spent between the ball bearing being released by the electromagnet and breaking the connection at the trap. This gives you a timing for the descent. You need to measure the distance of the drop. You should take at least four readings and average the result.

## The Simulation

The timer can be reset by clicking on the central red button. The circuit breaker can be moved by placing the cursor over the red handle and using the mouse wheel. You can replace the ball bearing on the electromagnet by clicking on the ball bearing. If you have not put the switch back to its 'on' position, the ball bearing will fall right away. The vertical distance fallen is 0.67 m.

You can repeat the experiment on the Moon by clicking on the Moon icon at the top left of the screen (see Figure 4.2), and on Mars by clicking on the Mars icon at the top left of the screen (see Figure 4.3).

**FIGURE 4.2**
The gravity apparatus on the Moon.

**FIGURE 4.3**
The gravity apparatus on Mars.

## The Results

You should prepare a table for your results that shows the timings taken and the average. You can then compute the rate of acceleration for the Earth or for the other worlds using the formula:

$$g = 2s/t^2$$

Where $s$ is the distance, $g$ is the acceleration (the value you are looking for), and $t$ is the time of descent.

If you are using the simulator, you can repeat this experiment for the Moon and Mars.

## Further Discussion

Look up the masses of these worlds on the internet. You could try plotting the masses against the values for $f$ that you found. Then answer the following questions:

- What is the relationship between the mass of a planet and its acceleration due to its gravity?
- Why is the graph not a straight line?
- What other factor might influence the value of the acceleration (hint: are all of these worlds the same size?)?

Is it possible to show that the Moon falls towards the Earth the distance we would expect when compared to an apple at the surface of the Earth?

In fact, this is surprisingly easy to show.

Here on Earth, the apple falls 4.9 meters in the first second. The Moon is 60 times farther away from the center of the Earth than the apple, so we should expect it to fall 1/60×60; i.e., 1/3600 of this, i.e., about 1.3 mm.

The angle the Moon moves through in 1 second is given by:

$$\Theta = 2 \times pi/(28 \times 24 \times 60 \times 60) \text{ radians}$$

The denominator is the number of seconds in a lunar month, which corresponds to a single orbit around the Earth.

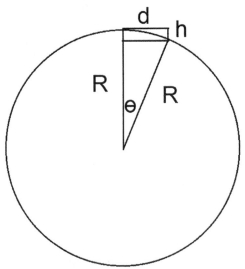

**FIGURE 4.4**
The geometry of the Moon's orbit, showing the amount the Moon falls in a small part of its orbit.

This works out at 2.59722E-06 radians. Given that the radius of the Moon's orbit is approximately 384,400,000 m, the distance moved by the Moon in 1 second is:

$$D = 2.59722\text{E-}06 \times 384,400,000 \text{ m} = 998.37 \text{ m}$$

From the earlier diagram:

$$R - h = R\cos(\theta)$$

Therefore, $h$, the amount that the Moon has dropped towards us, is given by:

$$h = R - R\cos(\theta)$$

Which works out at around 1.3 mm.

---

## Note

1. Cohen, I. B. *Isaac Newton's Papers and Letters on Natural Philosophy and Related Documents* (Cambridge, UK: Trinity College Library, 1958), 189.R.4.47, ff. 7–8.

# 5

# Average Velocity Using an AirTrack

## Introduction

The AirTrack is a very useful device for performing a variety of experiments about motion, forces, and momentum. What makes it so effective is its use of pumped air to create a virtually frictionless track upon which various glider configurations can be put in motion and collided against.

Figure 5.1 shows you the main components, where you can see the main track running from left to right with many small holes along its length. Air is pumped through these holes, which creates a virtually friction-free surface for the glider to travel along. There are two sets of support legs. The left-hand supporting legs can be raised and lowered. If your AirTrack does not have an adjuster, then blocks can be placed under the legs when you need to raise it. The buffers at the ends of the track have rubber bands that reverse the direction of the glider's velocity. Notice that there is a ruler alongside the track with a moveable pointer that can be used to reset the position of the glider. There is also a ruler at the left of the track for measuring how far it has been raised.

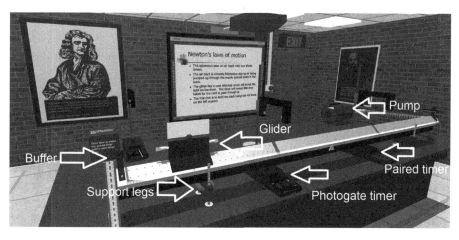

**FIGURE 5.1**
The components of the AirTrack.

DOI: 10.1201/9781003262350-6

The photogate timers are used to time the gliders traveling through them. The photogate timer has a light and a detector inside the U-shaped gantry. When the glider's white rectangular card interrupts the beam, the timer runs. The time that is recorded can be used to calculate the speed, as the length of the card is known.

This chapter and the one following are not full experiments but are exercises that allow you to become familiar with the workings of an AirTrack. The first exercise will show you how to level the track using the independent timers. The second exercise will show you how to measure average velocity using paired timers. Having done these exercises, you will then be able to go on and do experiments with some confidence and familiarity.

## The Objective

To level the AirTrack and determine the average velocity of a glider when in motion on an AirTrack.

## The Apparatus

You will need:

- An AirTrack
- An air pump
- A glider
- Two photogate timers, paired and not paired (when paired, the timers act together and record the time between entering the first gate and the second gate; when not paired, a photogate records the time that the glider takes to pass through this timer only)

The track should be configured as shown in Figure 5.3. For the first part of the exercise, the track should start off not being level.

## The Variables

The dependent variables are the time taken to travel between the paired photogate timers and the time to pass through an independent photogate timer.

## The Physics

The AirTrack will be level when the glider takes the same time to go through two photogate timers some distance apart on the track. This is clearly

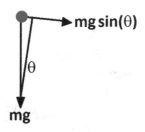

**FIGURE 5.2**
The forces on the glider.

indicating that the glider is neither accelerating nor decelerating due to gravity, which is a consequence of being level.

Average speed is simply calculated by dividing the distance between the photogate timers by the time it takes to travel between them.

If you wish to calculate the acceleration due to gravity, note that the force downwards on the glider is its mass times *g*. Resolving in the direction of motion, the force is *mg sin(θ)*.

## The Method

First, we are going to level the AirTrack. Using two photogate timers, put them at a reasonable distance apart, as shown in Figure 5.3.

Switch the photogate timers on. Start the blower and time the glider through both gates. If the glider goes through the second gate in a shorter time than the first, this indicates that it is accelerating, so lower the track at the left support and repeat. Keep doing this until you get the same times. When you have the track as level as possible, carefully measure the height at the left edge so that you can quickly find the level position again.

Measure and make a note of the length of the glider's fin.

Raise the left-hand side of the track a little so that there is an appreciable acceleration of the glider when the pump is on.

Move the position pointer to the middle of the scale.

Position the two photogate timers at equal distances to the left and right of the pointer, as far from each other as possible. The main photogate timer should be on the left or higher part of the track.

Raise the left-hand side of the track a little so that there is an appreciable acceleration of the glider when the pump is on.

Turn on the pair function of the timers so that the timer gates time how long the glider takes to pass between them.

Bring the glider to the left of the main photogate timer and note its position from the scale (you need to record this, as you will need to return the glider to this position for subsequent runs). Record the time taken to pass between the two gates and the distance between them.

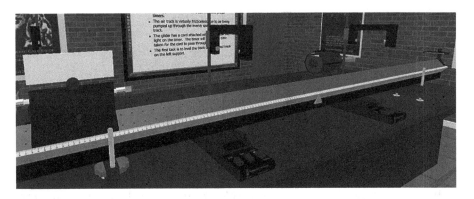

**FIGURE 5.3**
The placement of the photogate timers.

Repeat this for smaller and smaller distances that the timers are apart. Make sure they are always equidistant from the pointer. When the glider is to the left of your start position, you may find it useful to switch off the pump; that will stop it moving and give you time to rearrange the photogate timers to their new positions.

## The Simulation

You can swap the paired timers for two independent timers by clicking on the red cube on the back left of the table.

The left-hand side of the track can be raised or lowered by placing the mouse over the wheel and using the scroll wheel.

The green pointer can be moved left or right by putting the mouse cursor over it and then using the mouse wheel. You will use this to indicate the position between the timers so that they can be placed at an equal distance to the left and right of the pointer. Note that you can estimate the position of each timer accurately by examining its shadow on the track.

You can turn on the pair function of the timers by clicking the *Pair* button of the main timer.

The electronic photogate timer can be switched on by clicking the left-hand button.

The timer can be reset by clicking the middle button. It can act in paired mode by clicking the right button.

The pump can be turned on by clicking the red knob, as indicated in Figure 5.8.

There is a visible indication that the pump is on, as the connecting pipe will sway under the pressure of the air. The glider will only move freely when air is being pumped into the track.

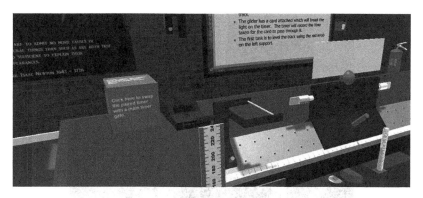

**FIGURE 5.4**
The scale showing the height of the AirTrack at one end.

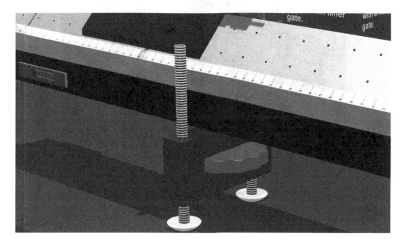

**FIGURE 5.5**
The height adjuster of the AirTrack.

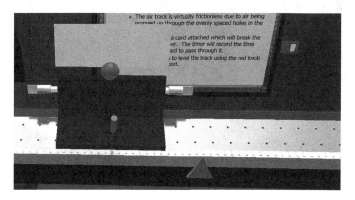

**FIGURE 5.6**
The position pointer.

**FIGURE 5.7**
The controls on the photogate timer.

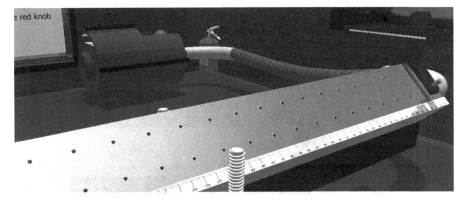

**FIGURE 5.8**
The air pump.

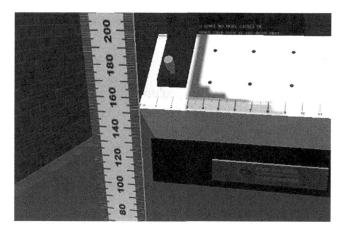

**FIGURE 5.9**
The ruler showing the height of the track at one end.

You can measure the height of the left edge of the track by using the ruler that is permanently placed there. See Figure 5.9.

In the simulation, you can move the glider by giving it an impulse using the mouse wheel when the mouse is over the red sphere between the body of the glider and the rectangular card.

You can hold the glider with the left mouse button and let go by releasing it.

## The Results

Record the time taken to pass between the two photogate timers and the distance between them in a table like this:

| Distance/cm | Time/s | V/cm s−1 |
|---|---|---|
| | | |
| | | |
| | | |
| | | |
| | | |
| | | |
| | | |
| | | |
| | | |
| | | |

1. Calculate the speed for each row and put the results in the third column.

2. Plot the speed against the distance. Mark the distance on the X-axis and the velocity on the Y-axis.

3. Why are the results in the third column different from each other?

4. What is the trend for the third column as the distances get shorter? Can you offer an explanation for this?

5. Which result gives the most accurate reading for the glider's velocity as it goes over the red marker?

6. Can you deduce what the actual instantaneous speed of the glider is at the midpoint of the graph?

7. Would this be accurate?

8. How might you improve the accuracy?

9. What is the limiting factor of the apparatus that prevents us from gaining a better result for the actual instantaneous speed of the glider at the midpoint?

## Further Discussion

When the track is leveled using the procedure described here and given that there must be some remaining friction in the track or from the air, the level found isn't strictly speaking the same level that might be found using a builder's level – it has also taken account of any remaining friction. Do you think this makes experimental results more accurate or less accurate? Is there a scenario that should be avoided when making measurements with a track that has been leveled in this way?

# 6

# *Determining Acceleration Using an AirTrack*

## Introduction

As was noted in the previous chapter, it is possible to do many experiments and measurements with the AirTrack. In this chapter, we are going to measure acceleration. This is not a full experiment but is intended to show you how such a measurement is made. In this configuration, we will simply raise one end of the track and let gravity provide the force that will accelerate the glider along the track. The next chapter will be concerned with an actual experiment that confirms Newton's second law.

## The Objective

To measure the acceleration of the glider on an AirTrack when subjected to a constant force from gravity.

## The Apparatus

You will need:

- An AirTrack
- An air pump
- A glider
- Two photogate timers and a paired gate

The AirTrack is assembled as shown in Figure 6.1. It does not need to be level.

## The Variables

The dependent variable is the time taken to travel between each of the two paired photogate timers.

DOI: 10.1201/9781003262350-7

## The Physics

Average speed through each timer gate is simply calculated by dividing the length of the card on the glider by the time it takes to travel through each timer gate. This gives two speeds: $V_1$ and $V_2$. The time it takes to travel from one timer gate to another is $Dt$. The acceleration is given by the change in velocity over the time $Dt$:

$$A = (V2 - V1)/Dt$$

If we take as an assumption that the acceleration is constant, then this result will be accurate. We can attempt to verify this by making the distance between the gates smaller and smaller and examining the calculated acceleration for each.

## The Method

Measure and make a note of the length of the glider's fin. Move the position indicator on the track to about one-fifth of the way along from the left. Raise the left part of the track (the opposite end to the pump), using the height adjuster, so that there is an appreciable acceleration when the pump is on.

Make sure the photogate timers are in paired mode, with the glider at the position of the indicator as shown in Figure 6.1.

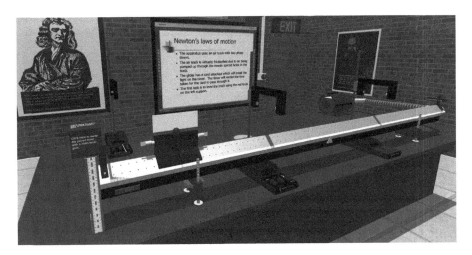

**FIGURE 6.1**
Configuration of the AirTrack for measuring acceleration.

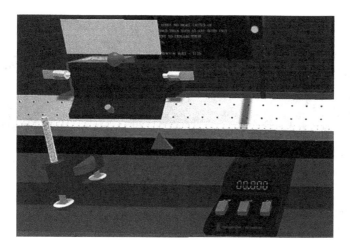

**FIGURE 6.2**
Position of glider at the pointer indicator.

Switch on the pump (or with the pump on, hold it there and then release). Make a note of how long it took the glider to travel between the two gates; this is *Dt*. You may want to stop the pump at this point.

Swap to using two independent timer gates in exactly the same positions as the two paired gates. Make sure these are both switched on and are not paired.

Take the glider back to the same position and, with the pump on, release it. Take note of the times through each gate; these are $T_1$ and $T_2$.

## The Simulation

The length of the glider's fin is 16.5 cm.

You can change from paired timer gates, which are used to measure the time of the glider passing between both, and non-paired timer gates that will independently measure the time taken for the glider to move through each, by clicking on the large red cube. Note that when the timer gates have been swapped, the new gate goes to the exact same position as where the old gate was. This can be convenient when taking a series of measurements that involve both sorts of timer gate.

The glider can be stopped by holding down the mouse button when the cursor is over the fin (the rectangular upper part). It can be released by releasing the mouse button.

An impulse can be applied to the glider by placing the cursor over the rectangular area and using the mouse wheel. This can be useful when you want to place the glider at a particular position on the track. Even with the pump off, it is possible to move the glider downhill by giving it an impulse. This is very useful when you want to position the glider accurately.

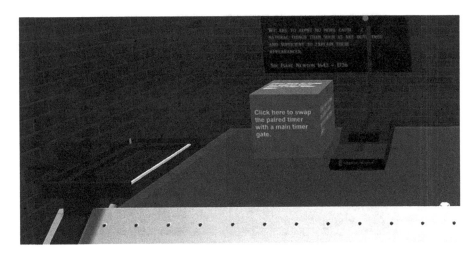

**FIGURE 6.3**
Switching the photogates.

The green pointer can be moved left or right by putting the mouse cursor over it and then using the mouse wheel.

The two timers can be moved left or right by putting the mouse cursor over their bases and using the mouse wheel.

The exact position of a timer can be accurately estimated from its shadow on the AirTrack's scale.

## The Results

Take repeated readings with different distances between the photogate timers, always starting the glider from the same position. Complete the table.

| Dt/s | T1/s | T2/s | V1/ ms⁻¹ | V2 ms⁻¹ | A/ms⁻² |
|------|------|------|----------|---------|--------|
|      |      |      |          |         |        |
| 9.76 |      |      |          |         |        |
|      |      |      |          |         |        |
|      |      |      |          |         |        |
|      |      |      |          |         |        |
|      |      |      |          |         |        |

Where $Dt$ is the time taken to travel between the gates (measured using the paired gates). $T1$ is the time through the first gate. $T2$ is the time through the second gate. *V1* is the average speed through the first gate. *V2* is the average speed through the second gate. *A* is the acceleration.

## Further Discussion

What happens if you start the glider from a position farther from the first photogate timer? Can you explain the result?

By noting the amount that the track has been raised, you can calculate the angle of the track. You can use this to calculate a value for *g*, the acceleration due to gravity.

Simulation users beware: the gravity in the simulation is about 15% of the Earth's!

If you tried to calculate *g* using a sloped ramp and a cylinder that rolled down the ramp, why would it not give a good result? Think about where all the potential energy is going to as the cylinder *rolls* down the slope.

# 7

## Confirmation of Newton's Second Law

### Introduction

Isaac Newton's work caused a revolution in how we looked at the world. He was without doubt a genius, teaching himself the mathematics known at the time and inventing calculus for good measure, because he needed it to qualify and quantify those things that involve changes. He worked mostly alone, although in later life he was Master of the Mint and president of the Royal Society. He was extremely secretive and often did not publish his work until many years after he had completed it. His lifelong interest in alchemy seems to be totally at odds with his extraordinary powers of lateral thought, observation, and mathematical insightfulness. He was the first scientist to receive a state funeral, and he left a legacy that the world and the universe beyond were ruled by the certainty of mathematics applied to simple but universal physical laws. His monument in Westminster Abbey, where he is buried, reads:

> Here is buried Isaac Newton, Knight, who by a strength of mind almost divine, and mathematical principles peculiarly his own, explored the course and figures of the planets, the paths of comets, the tides of the sea, the dissimilarities in rays of light, and, what no other scholar has previously imagined, the properties of the colours thus produced . . .

Newton's second law is perhaps the most well-known principle in Physics. Galileo came close to finding it, but it was Newton who finally nailed it as the force, $F$, acting on a body of mass $m$, causes an acceleration, $a$, where they are related by:

$$F = ma$$

This is something that we can confirm with the AirTrack by applying a series of different forces to the glider and measuring its acceleration using the photogate timers.

DOI: 10.1201/9781003262350-8

## The Objective

To confirm Newton's second law that force = mass times acceleration.

## The Apparatus

You will need:

- An AirTrack and pump
- A glider
- Two pulleys and holders
- A thread
- A set of weights, 10 g–100 g
- Two non-paired photogate timers

The apparatus is set up as shown in Figure 7.1. Two of the pulleys can be suspended from the ceiling so that a large drop for the weight is available.

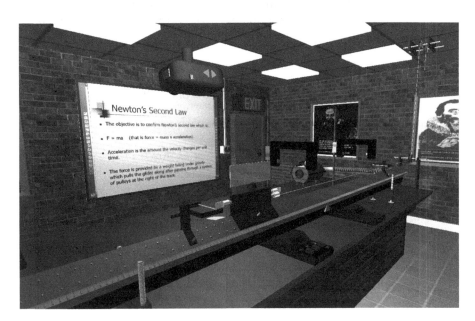

**FIGURE 7.1**
AirTrack for Newton's second law.

## The Variables

The independent variable is the mass controlling the force on the glider. The dependent variable is the acceleration of the glider.

## The Physics

The acceleration is given by the change in velocity. $Dt$ is the time to travel from the first gate to the second. $V_1$ is the average speed through the first timer gate; $V_2$ is the average speed through the second timer gate. As in the previous chapter, we calculate the acceleration by the change in speed over the time $Dt$:

$$A = (V_2 - V_1)/Dt$$

If we take as an assumption that the acceleration is constant, then this result will be accurate. We can verify this by making the distance between the gates smaller and smaller and examining the calculated acceleration for each.

## The Method

With the glider to the left of the first gate, mark its position so that repeat runs can be made from the same place. Start by using the lightest weight.

**FIGURE 7.2**
The non-paired timer gates.

Using the paired timer gates, time how long it takes for the glider to go from the first timer to the second. Note this time as *Dt*.

Swap the timers for two non-paired timers. Let the glider go from the same position and note the time it takes to go through both gates. Note these times as *t1* and *t2*.

Repeat the process with increasing weights. For each weight, you can do several runs so that the results can be averaged.

## The Simulation

The descriptions of the various controls are the same as in Chapter 6, with the addition of being able to add weights to the cradle by placing the mouse cursor over the cradle and using the mouse wheel. The weights are 10 g, 20 g, and 30 g.

## The Results

Complete the table:

| Mass/g | t1/s | V1/cm s⁻¹ | Dt/s | t2/s | V2/cm s⁻¹ | a/cm s⁻² |
|--------|------|-----------|------|------|-----------|----------|
|        |      |           |      |      |           |          |
| 10     |      |           |      |      |           |          |
| 20     |      |           |      |      |           |          |
| 30     |      |           |      |      |           |          |
|        |      |           |      |      |           |          |

Your weights may be different, and you may have more of them. Add more rows if you have done repeated runs, and calculate the average of each repeated set.

Plot the weight (Mass X g) on the X-axis and the acceleration on the Y-axis, and you should obtain a straight line.

## Further Discussion

Explain why the best-fit straight line does not go through the origin.

# 8

## Showing Conservation of Energy Using an AirTrack

### Introduction

This uses the exact same setup as the chapter showing how to measure speed using an AirTrack (see Chapter 6). But this time we will use it to show that as the glider gains speed, and hence kinetic energy, it loses potential energy as its height decreases. With some careful measurements, you should be able to show that the energy gained by the glider speeding up is the same as the energy lost by height.

### The Objective

To establish conservation of energy in the case of an object gathering speed under the force of gravity.

### The Apparatus

You will need:

- An AirTrack
- An air pump
- A glider
- Two photogate timers (paired and non-paired)

The AirTrack is configured as shown in Figure 8.1. There is no need for the track to be level.

### The Variables

The independent variable is the height lost. The dependent variables are the time taken to travel between the two paired photogate timers and the time taken to go through each gate.

DOI: 10.1201/9781003262350-9

**FIGURE 8.1**
AirTrack with paired timer gates.

## The Physics

The potential energy of an object due to gravity is given by:

$$PE = mgh$$

Where the mass of the object is $m$, the acceleration due to gravity is $g$, and the height is $h$. You can think of this in terms of the work needed to be done in raising a mass by a height $h$. Work done is calculated by the formula:

$$W = Fd$$

That is, the work done is equal to the force multiplied by the distance over which the force is applied. In this case, the force is $mg$ and the distance is $h$.

The kinetic energy of an object is given by:

$$KE = mV^2/2$$

Where $V$ is the object's speed. We will be equating the kinetic energy gained to the potential energy lost.

## The Method

Raise the left side of the AirTrack by a few centimeters. Do not raise it so high that the glider moves without the pump being switched on.

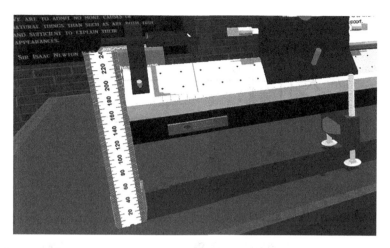

**FIGURE 8.2**
Feet assembly with height adjustment and ruler.

With the paired photogate timers at a good distance from each other, as shown in Figure 8.1, mark a starting point for the glider before the first gate. Move the glider to this position and, with the pump on, release it. Note the time it takes to go from the first gate to the second as *Dt*.

Now, using two independent photogate timers, release the glider from the same position and note the time through the first gate as $T_1$ and the time through the second gate as $T_2$.

Measure the height of the track at the first gate and note this as $h_1$; then measure the height of the track at the second gate and note this as $h_2$. Calculate *Dh* as the difference between the two heights.

$$Dh = h_1 - h_2$$

Repeat this for different heights of the track.

## The Simulation

The track can be raised on the left side by using the wheel on the feet assembly. A ruler at the farthest end of the track can be used to measure by how much it has been raised. Make a note of this value as *DTh* (delta track height).

The amount the track is raised at the first and second gates can be calculated from the amount you raised the track at its start, *DTh*, and the position of the gate. The second support is at 1.78 m. The geometry is given in Figure 8.3.

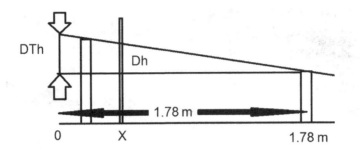

**FIGURE 8.3**
The AirTrack's geometry.

**FIGURE 8.4**
Using shadow to position a photogate.

The height raised at any position X is therefore given by:

$$Dh = (1.78 - X) \, DTh \, / \, 1.78$$

You can immediately verify this by putting *X = 0*, which gives *Dh = DTh*. This is clearly true, as this is the amount that we measured the leftmost part of the track as being raised. Now put *X = 1.78* and *Dh = 0*. This is the height of the AirTrack above its rightmost support, which is not raised at all, so this is correct too.

Use the formula to calculate the value *Dh* from both gate positions. Note that the gate positions can be observed by using their shadows from the overhead lighting on the AirTrack scale.

Note that the acceleration due to gravity in the simulation is a good deal less than that found on Earth (*0.153 * g*).

## The Results

Complete the table with the results that you find for different raised amounts.

| Dt/s | T1/s | T2/s | V1/ms⁻¹ | V2/ms⁻¹ | KE1/J | KE2/J | DKE/J |
|------|------|------|---------|---------|-------|-------|-------|
| 6.62 | 2.019 | 0.879 | 0.081724 | 0.187713 | 0.00167 | 0.008809 | 0.007139 |
|  |  |  |  |  |  |  |  |
|  |  |  |  |  |  |  |  |
|  |  |  |  |  |  |  |  |
|  |  |  |  |  |  |  |  |

This tabulates the data for the kinetic energies gained. You can also create a table for the potential energies lost (or combine both tables, which would be difficult here due to the needed width but can be done easily in a spreadsheet).

| GP1/m | GP2/m | DTh/m | h1/m | h2/m | Dh/m | Mgh/J |
|-------|-------|-------|------|------|------|-------|
| 0.548 | 1.388 | 0.02 | 0.013843 | 0.004404 | 0.009438 | 0.007083 |
|  |  |  |  |  |  |  |
|  |  |  |  |  |  |  |
|  |  |  |  |  |  |  |
|  |  |  |  |  |  |  |

The values given here are from the simulator with the track raised by 2 cm. The last value in the rows gives the kinetic energy gained and the potential energy lost, which you can see are quite close in value.

## Further Discussion

A roller coaster is a little bit like the AirTrack but without the friction-reducing measures, and it generally uses wheels. However, the conversion from potential energy to kinetic energy and vice versa happens multiple times as the track weaves up and down. Can you list all the sources of friction in this scenario? What happens to the energy lost through friction?

# 9

## Conservation of Momentum in an Inelastic Collision Using an AirTrack

### Introduction

In an inelastic collision, the bodies collide and then stick together. In this sort of collision, there will be a loss of kinetic energy that will sometimes manifest itself by work done or heat generated. In the case of a car accident, for example, this could be the work done on bending metal and breaking glass.

### Objective

To confirm the conservation of momentum for inelastic collisions using an AirTrack and two gliders of various weights.

### The Apparatus

You will need:

- An AirTrack
- Two independent photogate timers
- Two gliders with Velcro pad bumpers
- A set of weights for the gliders

The AirTrack is assembled as shown in Figure 9.1. Note the use of two gliders at the same time.

### The Variables

The independent variables are the masses of the gliders and the velocity of the first glider. The dependent variable is the velocity of both the gliders after the collision.

DOI: 10.1201/9781003262350-10

## The Physics

Momentum of a body is given by the mass of a body times its velocity, which makes it a vector.

$$\mathbf{p} = m\mathbf{v}$$

The conservation of momentum is a direct consequence of Newton's second law, which can be stated in terms of momentum as force is equal to the rate of change in momentum, or:

$$\mathbf{F} = D\mathbf{p}/Dt$$

In a system where there are no forces acting, there will be no change in momentum; therefore, momentum must be conserved.

## The Method

Level the track if necessary. With the two photogate timers at a reasonable distance apart, position the left glider before the first photogate timer and the right glider just to the left of the second photogate timer, as shown in Figure 9.1.

With the blower turned on, give an impulse to the first glider; make sure that you have finished applying force to the glider before it enters the first

**FIGURE 9.1**
AirTrack with two gliders.

photogate timer. Take note of the time it takes for the first glider to pass through the left-hand photogate timer as $T_1$. The two gliders will then collide and stick together. Time the second glider through the right-hand photogate timer; this is $T_2$. Be careful to note the time through the second photogate timer before the first glider reaches it. You can either stop the gliders with your hand or just make the observation quickly, whichever is more convenient.

This should be repeated for different masses on both gliders.

## The Simulation

Use two independent photogate timers by clicking on the red cube at the back left of the table. The left glider will need to be to the left of the left-hand gate, and the right glider will need to be to the right of the right-hand gate, before you are allowed to swap them. In addition, the blower needs to be switched off. The gliders can be positioned by applying an impulse using the mouse wheel when the cursor is over the glider. The glider can be stopped when in motion by clicking on it.

The blower can be turned on and off by clicking on its red button.

The glider weight is 350 g in its default configuration of having no extra weights added. You can change the mass on the glider by hovering the mouse cursor over the glider and using the mouse wheel. This must be done when the blower is turned off.

Note that weights are added in pairs of 25 g; one goes on each side of the glider. You can add up to four pairs, making the glider have a maximum total mass of 550 g. When you add a weight on one side of the glider, another is added on the other side automatically.

The length of the glider's fin is 16.5 cm.

You can reset the timers by clicking on the middle red button.

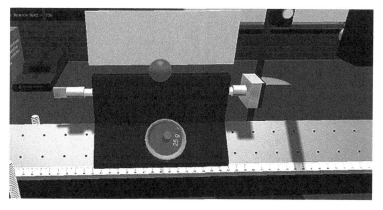

**FIGURE 9.2**
Weights added to a glider.

## The Results

Complete the following table using your own results:

| M₁/kg | M₂/kg | T1/s | T2/s | V₁/ms⁻¹ | V₂/ ms⁻¹ | Pᵢ | Pₐ |
|---|---|---|---|---|---|---|---|
| | | | | | | | |
| 0.35 | 0.35 | 2.398 | 4.799 | 0.069 | 0.034 | 0.024 | 0.024 |
| 0.4 | 0.35 | 1.079 | 2.059 | 0.153 | 0.080 | 0.061 | 0.060 |
| 0.45 | 0.35 | 1.121 | 1.999 | 0.147 | 0.083 | 0.066 | 0.066 |
| | | | | | | | |
| | | | | | | | |
| | | | | | | | |

$V_1$ and $V_2$ are calculated from the length of the card divided by *T1* and *T2*, respectively. $P_i$ is $M_1{}^*V_1$, *Pa* is $M_1{}^*V_2 + M_2{}^*V_2$. The first few rows show results from the simulation; these show a good correlation between the initial and after momentums.

## Further Discussion

Can you think of a way of repeating the experiment with the second glider in motion, with a known velocity before the collision?

# 10

## Hooke's Law

### Introduction

Robert Hooke was a contemporary of Isaac Newton, Christopher Wren, and Robert Boyle. He engineered the vacuum pumps for Boyle's experiments with gases. He achieved in many scientific areas mainly as the curator for experiments at the Royal Society, but he was also the Surveyor to the City of London and performed many of the surveys necessary after the Great Fire of London in 1666. The law that bears his name and concerns the relationship between the force on an elastic substance and its extension was discovered by Hooke in 1660, although it was not published until some eighteen years later. He put the result into practice by inventing the hairspring – a device that made the portable timepiece or pocket watch possible.

When we apply a force to stretch something and, on letting go, it returns to its original size and shape, we say that the object has undergone an elastic deformation. These are the kind of deformations that we are concerned with here. When we stretch elastic or a steel wire with some force, it will stretch, and intuitively we are aware that the bigger the force, the more it will stretch. We are also aware that some materials stretch more than others. What exactly is the relationship between the force and the amount something stretches for a particular material? That is the question that we are going to answer with this experiment.

Hooke's law states that the amount that an elastic substance stretches (the extension) is proportional to the force applied. Thus:

$$F = ke$$

Where $F$ is the force, $e$ is the extension, and $k$ is the constant of proportionality. This is known as the stiffness constant, and it is different for different materials.

### The Objective

To show that the amount that an elastic wire stretches (the extension) is proportional to the force applied.

DOI: 10.1201/9781003262350-11

## The Apparatus

You will need:

- A ruler
- A protractor
- A thin steel wire (a top E guitar string works well)
- A set of weights
- A weight cradle
- Two G-clamps
- An axle with pulley and indicator
- Axle supporting block
- A wood block

The thin wire is held at one end between a block of wood and a G-clamp. For extra firmness, it can be looped and pinned to the left of the G-clamp, as shown in Figure 10.1. The pulley has a pointer attached, the angle of which can be measured from the protractor behind it. The weight applied can be changed by increasing the number of weights in the cradle.

## The Variables

The independent variable is the weight, and the dependent variable is the angle.

**FIGURE 10.1**
Hooke's experiment apparatus.

## The Physics

To convert Theta (or $\Theta$, if you'd like to use the Greek letter for theta, which is much more sophisticated) degrees to radians, you can multiply by $2\pi/360$. Your calculator may have an immediate way of doing this.

To calculate the extension, delta $x$ (or $\Delta x$), use the formula:

$$\Delta x = \Theta r$$

Where $r$ is the radius of the spindle.

## The Method

Start with a weight that makes the wire taut without stretching it very much. Take note of the angle the pointer indicates, or if you can reset the pointer to zero, then do this. For each possible additional weight, observe the further angle of rotation of the pointer on the protractor.

## The Simulator

In the simulator, we are assuming that the cradle is supplying sufficient force to minimally extend the wire. We are taking this as the zero point for the force and the extension. It is at this cradle-weight that we are measuring the angle from.

**FIGURE 10.2**
Adding weights to increase the tension in the wire.

The weights in the cradle can be altered by placing the mouse cursor over the cradle and using the mouse wheel or P/N.

Note that the radius of the pulley's spindle is 5 mm, and the unstretched length of the wire to the pulley is 80 cm.

## The Results

For each possible weight, calculate the extension of the wire. The amount of weight can be changed by putting the mouse cursor over the weight cradle and rotating the mouse wheel. Create a table with a column for the weight and a column for the extension, like this but with your own results:

| M/kg | Theta/deg | ΔTheta/deg | ΔTheta/rad | Delta x/m | Force/N |
|------|-----------|------------|------------|-----------|---------|
| 0 | 270 | 0 | 0 | 0.00E+00 | 0 |
| 1 | 287 | 17 | 0.296556 | 1.48E-03 | 9.81 |
| 2 | | | | | |
| 3 | | | | | |
| 4 | | | | | |

Plot a graph of force against extension and verify that this is a straight line that demonstrates that the extension is proportional to the force.

Calculate the stiffness constant from the formula given at the start of this chapter.

## Further Discussion

In this version of the experiment, the wire is wrapped around a spindle, which has the pointer that records the angle the spindle has rotated. This effectively amplifies the extension, making it easy to measure very small extensions. However, wrapping the wire tightly around a spindle is likely to introduce some unwanted friction into the system. Can you devise a method by which this effect can be quantified?

# 11

## Young's Modulus

## Introduction

Born some seventy years after Robert Hooke's death, Thomas Young refined Hooke's elasticity law by identifying the modulus described in this chapter that is independent of the dimensions of the material, thus aiding engineers in their calculations. Young made extensive contributions to many other areas of science – in particular, optics, in which the double-slit experiment bears his name.

The problem with a stiffness constant, which we measured in Chapter 10, is that its value will depend on the dimensions of the material used. If we want to be able to compare materials, we need a measure that is independent of the materials' dimensions. This requires us to use stress and strain rather than force and extension.

We need to define these terms accurately, which we will do in the Physics section of this chapter. They are known formally as tensile stress and tensile strain (tensile just means something to do with stretching).

## The Objective

To measure Young's modulus for a thin wire.

## The Apparatus

See the list of apparatus in Chapter 10.

## The Variables

The independent variable is the weight from which the tensile stress is calculated, and the dependent variable is the angle from which the strain is calculated.

DOI: 10.1201/9781003262350-12

## The Physics

Tensile stress is defined as the force applied divided by the cross-sectional area. This can be expressed as:

$$\text{Stress} = F/A$$

Where $F$ is the force applied to stretch the wire and $A$ is the cross-sectional area of the wire. The units are $Nm^{-2}$, which is the same as what we measure pressure with, i.e., Pascals or Pa.

Tensile strain is defined as the change in length as a proportion of the original length. This can be expressed as:

$$\text{Strain} = e/l$$

Where $e$ is the amount that the wire has stretched and $l$ is its original length. This is a ratio of two numbers with the same units. So the units are $mm^{-1}$, which cancel, so there are no units at all. Strain is simply a numeric ratio.

Young's modulus is defined as the stress divided by the strain. That is:

$$E = \text{tensile stress/tensile strain}$$

Substituting from the previous two definitions, this gives us:

$$E = Fl/Ae \text{ (units Pa)}$$

If we take a wire and stretch it within its elastic limit, it will return to its original length when we stop applying a stretching force. However, if we were to carry on applying more and more force, there comes a point – the elastic limit – that when exceeded no longer allows the wire to go back to its original length. Any deformation beyond this elastic limit is called a plastic deformation. We are not concerned with such deformations here.

## The Method

The method is identical to that used in demonstrating Hooke's law, described in Chapter 10.

## The Simulator

See the notes in Chapter 10.

## The Results

Create a table with headings for the mass, the angle of deflection (call this theta) in degrees, the same angle in radians, the amount of stretch (call this delta x), the force, the stress, and the strain. Your table should look something like this:

| M/kg | Theta/deg | ΔTheta/deg | ΔTheta/ rad | Δx/m | Force/N | Stress/ Nm⁻² | Strain (no units) |
|------|-----------|------------|-------------|------|---------|--------------|-------------------|
|      |           |            |             |      |         |              |                   |
| 0    | 270       | 0          | 0           | 0.00E+00 |     |              | 0.00E+00 |

You may find it convenient to do this using a spreadsheet program such as Microsoft Excel.

The stress and strain can be calculated from the formulae given in the Physics section.

To calculate the cross-sectional area of the wire, use the formula for the area of a circle:

$$A = \pi r^2$$

Once your table is complete, plot a graph with stress on the Y-axis and strain on the X-axis for all the readings that you have taken. You should get a straight line where the gradient equals the value for Young's modulus.

## Further Discussion

What do you think the area under the graph might represent? Think about what it is the product of.

What are the units on the two axes?
What do you get if you multiply these units?
What is force times a distance?
Is this something that you have met before?
Where else do you meet this formula?
What is measured by this formula?
Write down an equation for the strain energy in a stretched wire.

# 12

## Velocity of Rifle Shell Using a Ballistic Balance

### Introduction

Here a rifle or air pistol shoots a projectile at a suspended block of wood. The projectile buries itself into the wood inelastically and transfers its energy to the block; some of the energy is used up as noise, heat, and the effort needed to burrow into the wood. The block then moves sideways and upwards, converting its kinetic energy into potential energy. The height that the block rises indicates the amount of potential energy that the block and projectile combination gain. This can be equated to the kinetic energy of the block and projectile combination at the point when the projectile buries itself in the wood and the two start moving. This allows us to calculate the momentum

**FIGURE 12.1**
The rifle pointed at the ballistic balance's wood block.

DOI: 10.1201/9781003262350-13

of the block and projectile at this time, which due to the law of conservation of momentum must be equal to the momentum of the projectile before it strikes the block.

In the real world, for safety reasons, we usually perform this experiment with an air pistol at an aperture in a clear plastic cage, which encloses the ballistic balance. This ensures that no one can get hurt from a stray pellet.

Our simulation has no need to be safe and so uses an AK-47 assault rifle, as shown, which is most certainly not to be recommended for laboratory experiments!

## The Objective

To find the velocity of a projectile by using a ballistic balance.

## The Apparatus

You will need:

- A caged ballistic balance consisting of a suspended block of wood
- An air pistol
- Some pellets

The apparatus should be configured as shown in Figure 12.2. Care should be taken that the pistol is not able to fire outside of the cage.

## The Variables

The variables are the amount of horizontal deflection of the wooden block and the mass of the block with the projectile.

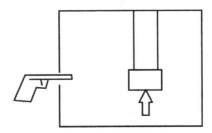

**FIGURE 12.2**
Schematic of the experiment.

## The Physics

The rope is attached at the top to point $P$. The following assumes that you are using a rope of length 1.5 m. If you are using a different length, then substitute the actual length in the following equations.

The end swings from $A$ to $B$. As $AP = BP$, $PAB$ is an isosceles triangle and the angles $PAB$ and $ABP$ are both the same, $\emptyset$.

The block moves horizontally by the distance that you measure in the experiment; we will call this $d$. Using trigonometry:

$$\sin(\Theta) = d/1.5$$

This enables you to calculate $\Theta$ and then $\emptyset$ (the total angle of the triangle $PAB$ must be 180 degrees). The height, $h$, is then given by:

$$h = d/\tan(\emptyset)$$

Now you can calculate the potential energy gained by the block. (Remember that the potential energy of a mass, $m$, at a height, $h$, with acceleration due to gravity given by $g$, is given by $mgh$.):

$$PE = (M_b + M_p)\, gh$$

Where $M_b$ is the mass of the block and $M_p$ is the mass of the projectile. The kinetic energy of the block at the point of impact is given by (remember that the kinetic energy of a mass, $m$, traveling at speed, $v$, is given by $mv^2/2$):

$$KE = (M_b + M_s)\, V_b^2/2$$

Where $V_b$ is the speed of the block. This value can now be equated to the kinetic energy of the block and projectile at the point of impact.

$$(M_b + M_p)\, gh = (M_b + M_p)V_b^2/2$$

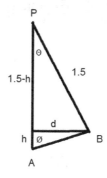

**FIGURE 12.3**
Geometry of the block's movement.

This can be rearranged to give a value of $V_b$, the velocity of the block and projectile just after the point of impact, i.e., when the bullet is no longer moving relative to the block.

$$V_b = \sqrt{2gh}$$

Now the Conservation of Momentum law can be used to equate the momentum of the projectile before it hits the block to the momentum of the bullet and block after the impact.

$$M_p V_p = (M_b + M_p) V_b$$

This can be rearranged to give $V_p$, the velocity of the projectile before it hits the block.

## The Method

You need to record the at-rest position on the horizontal scale. Note this initial position and then fire the gun. Note the farthest point of the block's swing. You may have to repeat this several times to get an accurate reading.

Once you have the reading for the position of the farthest point reached, you can then calculate the horizontal distance moved by the block by subtracting the at-rest position.

**FIGURE 12.4**
The wooden block's position scale.

## The Simulation

In the simulation, the length of the rope is 1.5 m. The mass of the wooden block, $M_b$, is 2 kg and the mass of the projectile, $M_p$, is 0.01075 kg.

You can fire the rifle by using the space key. You should only do this when the block is completely at rest. Note that if you use more shots, the combined weight of the block and the projectiles is going to need to take into account just how many times the rifle has been fired, as each projectile will add weight to the block.

To stop the block swinging, click on it. This will save you time, rather than just waiting for it to come to rest. You need to do this at the lowest point of the swing, and you may need to repeat the operation several times to bring the block to a halt. You can zoom in and out using +/- or the Z and X keys.

## The Results

Use the formulae in the Physics section to first calculate the potential energy gained by the block, and hence the velocity of the block just after impact.

The final equation in the Physics section can then be used to calculate the velocity of the projectile.

## Further Discussion

If a rifle like an AK-47 were to be fired when perfectly horizontal on a large flat landscape, and if the shell casing was to be ejected from the breech of the rifle horizontally, ignoring any effects from air resistance, which would hit the ground first, the projectile or the casing? Justify your answer.

# 13

## Simple Pendulum

## Introduction

A simple swinging pendulum can be used to estimate the acceleration due to gravity. The acceleration is given by the formula:

$$g = 4\pi l/T^2$$

Where $l$ is the length of the pendulum and $T$ is the period for an entire swing (i.e., back and forth).

If we take measurements of the period of the swing for a range of lengths and then plot $T^2$ against $l$, the gradient multiplied by $4\pi$ will be the acceleration due to gravity.

## The Objective

To find the acceleration due to gravity from the swing of a pendulum.

## The Apparatus

You will need:

- A pendulum bob
- A length of string
- A retort stand and clamp
- A G-clamp
- A meter ruler with its own stand
- A photogate timer or stopwatch

Clamp the retort stand to the bench at the base. Use the clamp on the stand to pinch the string at the required length. Attach the bob to the other end of the string.

There are three stands supporting various parts of the equipment and a U-gate timer. One of the stands supports the cord with a bob on the end.

DOI: 10.1201/9781003262350-14

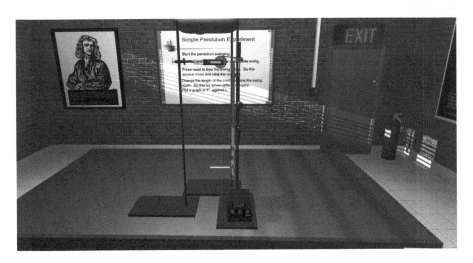

**FIGURE 13.1**
The experiment apparatus.

**FIGURE 13.2**
U-gate timer.

Another supports a ruler by which the length of the swinging part of the cord can be measured. If required, a third stand can support a bobbin for the cord. This can be omitted if the string is to be simply threaded through and pinched by the retort clamp.

At the bottom is the U-gate timer, which can be seen more clearly from a different angle, as shown in Figure 13.2.

Note that the pendulum length has been adjusted with the handle so that it swings through the U-gate, thus starting and stopping the timer when it swings.

---

## The Variables

The length of the string is the independent variable, and the dependent variable is the period of the swing.

---

## The Physics

Consider the position where the pendulum is partway through its swing:

Where it has moved in an arc $x$ from the equilibrium position due to a swing angle of $\theta$, the two forces acting on the bob are the tension in the string and the force downwards due to gravity. The restoring force, in the direction of a tangent to the arc, is:

$$F = -mg \sin(\theta)$$

Therefore, the equation of motion in the direction of the tangent is:

$$m\frac{d^2x}{dt^2} \quad mg \sin(\ )$$

We can divide both sides by $m$, and because $x = l\theta$, we can rewrite this as:

$$\frac{d^2}{dt^2} \quad \frac{g}{l}\sin(\ )$$

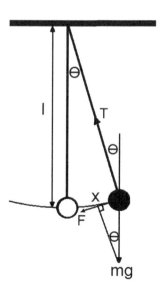

**FIGURE 13.3**
The forces acting on the pendulum.

For small angles, $\theta$ is a good approximation of $sin(\theta)$; therefore, the equation of motion is:

$$\frac{d^2}{dt^2} \quad \frac{g}{l}$$

This is the equation for simple harmonic motion, where the restoring force is proportional to the distance (or angle, as in this case) from the equilibrium position. For such motion, the period, $T$, is given by:

$$T = 2\pi\sqrt{(l/g)}$$

This can be squared, and then rearranged to give an equation for $g$ as a function of $T^2$ and $l$.

## The Method

Set the length of the cord by pinching it at its upper end with the clamp. Use the ruler to measure the length.

You can start the pendulum swinging by displacing it. Before taking any timings, you need to let the pendulum settle down from any random motions along its length. If there is not a sufficient swing, the timer gate will not function correctly, and you may get erroneous timings. If the pendulum is swinging too much, you may get nonharmonic motion that will give you inaccurate results.

Take three timings for each length and take the average. Make observations for at least six different lengths of the string.

## The Simulation

Note that this experiment is taking place in a laboratory on an unidentified planet. You should not assume that the acceleration due to gravity is the same as on Earth.

Set the length of the cord by placing the mouse over the bobbin winder and using the scroll wheel; this will wind the cord onto or off of the bobbin and thus change the length of the cord and the height of the bob.

You can adjust the position of the U-gate by using the mouse wheel, with the mouse over the U-gate's adjuster – the part that joins the gate to the vertical rod along which it slides (see Figure 13.5).

It is easier to get the position of the U-gate accurately set when the timer is switched on, as you can see when the light beam is interrupted. Switch the timer on by clicking on its leftmost button, as shown in Figure 13.6.

You can now read the length, $l$, by using the scale to the left of the bob. Note that there is a small metal pointer to help you do this accurately; see Figure 13.7.

**FIGURE 13.4**
Adjusting the length of the pendulum.

**FIGURE 13.5**
Adjusting the position of the U-gate.

**FIGURE 13.6**
Starting the timer.

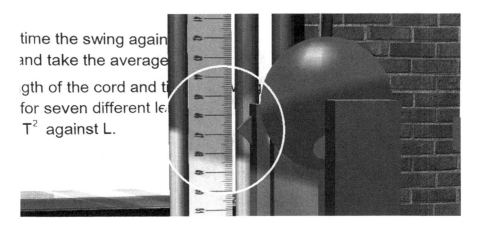

time the swing again
and take the average

gth of the cord and t
for seven different le
$T^2$ against L.

**FIGURE 13.7**
Measuring the length of the pendulum.

You can apply a force to the bob by placing the mouse pointer over the bob and rotating the mouse wheel. You can repeat this action until you get a sufficient swing.

## The Results

Create a table like this:

| L /m | T1 /s | T2/() | T3 /s | Tav /s | T² /s² |
|------|-------|-------|-------|--------|--------|
| 30 | 1.85 | 1.85 | 1.84 | 1.846667 | 3.410178 |
| 40 | | | | | |
| 50 | | | | | |
| 60 | | | | | |
| 70 | | | | | |
| 80 | | | | | |
| 90 | | | | | |

The table should include headings for the length, the three timings, the average, and the timing squared. When you have completed the table, plot the time squared against the length of the pendulum. Now compute the acceleration due to gravity using the gradient and the formula given in the first section.

## Further Discussion

The Physics section made the approximation that $\theta$ is a good approximation of $sin(\theta)$ for small angles, but just how small? Investigate this and find out how big the angle needs to be for the approximation to be inaccurate by more than 1 hundredth. Do you find the result surprising?

You need to express your angles in radians. For example $1° = 0.1753293$ and $sin(0.17453293) = 0.17452406$. This exercise can be done very efficiently on a spreadsheet.

# 14

## Simple Harmonic Motion Using a Mass-Spring System

### Introduction

A coil spring that is firmly attached at the upper end and with a weight attached to the lower end will oscillate up and down with simple harmonic motion (SHM) when given a small downward impulse.

The motion is called harmonic because of its association with the kinds of waves made by musical instruments. For example, the string of a guitar when plucked will move with SHM, and this motion is then transferred to the air as sound waves. Perhaps surprisingly, any wave can be represented by superimposing a number of different SHM waves. This was discovered by Jean-Baptiste Joseph Fourier while he was investigating the heat equation (how heat spreads), and it led to the mathematical process of Fourier analysis, which can be used to discover all the harmonic components of a wave.

### The Objective

To investigate how the time-period of a mass-spring system varies under different tensions.

### The Apparatus

You will need:

- A weight cradle and set of weights
- A coil/helical spring
- Length of string
- Retort stand and two clamps
- G-clamp
- A meter rule
- A timer

DOI: 10.1201/9781003262350-15

**FIGURE 14.1**
The experiment apparatus.

Clamp the stand to the table using the G-clamp around the base. Use the stand to support the top of the spring with a clamp. At the lower end of the spring, attach the weight cradle with a hook.

Use the second clamp, which can then be moved up and down, as a fiducial marker. The timer is used for timing the periods of the oscillations.

## The Variables

The weight on the spring is the independent variable. The period of the oscillations is the dependent variable.

## The Physics

Consider when the spring is stretched from its natural length by a weight and not in motion so that it is in equilibrium, as shown in Figure 14.2.

There is the force from the weight acting downwards and the force from the tension in the spring acting upwards. These are equal and opposite, and we can use Hooke's law to find the tension, so that:

$$T_1 = ke = mg \qquad \text{(eq1)}$$

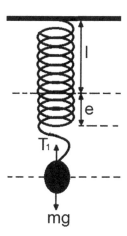

**FIGURE 14.2**
The forces acting on a mass-spring system when stationary.

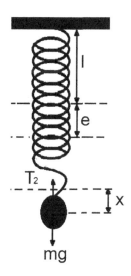

**FIGURE 14.3**
The forces acting on a mass-spring system when in motion.

Where $k$ is the spring constant (the constant of proportionality in Hooke's law) and $e$ is the extension. So an expression for $k$ is:

$$k = mg/e$$

When the spring and weight are in motion, we have the situation shown in Figure 14.3.

The spring is stretched by an extra amount $x$ and the force upwards is now $T_2$ given by:

$$T_2 = k(e + x)$$

So the net force is given by:

$$mg - T_2 = mg - k(e + x)$$

Using Newton's second law, this can be equated to the mass times the acceleration giving:

$$ma = mg - k(e + x)$$

From eq1, $ke = mg$, so we can simplify this equation to:

$$ma = -kx$$

This is the hallmark by which we know SHM; there is a restoring force proportional to the movement or displacement. We can rewrite the equation as:

$$a = -(k/m)x$$

Which can be compared to the normal equation for SHM, which is:

$$a = -\omega^2 x$$

Where:

$$\omega = \sqrt{(k/m)}$$

The period for SHM is given by:

$$T = 2\pi/\omega$$

Which in this case will be (we just need to put our equation for $\omega$ into this formula):

$$T = 2\pi\sqrt{(m/k)}$$

Thus, the theory of SHM tells us that the period is related to the mass and the spring constant by the formula:

$$T^2 = 4\pi2m/k$$

Where $k$ is the spring constant, $m$ is the mass, and $T$ is the period for an entire cycle (down from the midpoint, up through the midpoint, and down to the midpoint). This formula can be rearranged to find the spring constant:

$$k = 4\pi^2 m/T^2$$

If we take measurements of the period of the oscillation for a range of masses and then plot $T^2$ against $m$, then $k$ will be $4\pi^2$ divided by the gradient.

## The Method

Move the lower clamp so that it marks the position of the weight cradle, as shown in Figure 14.4.

Apply a downward impulse or two to the weight carrier. The spring system will start oscillating.

Start the timer as the carrier goes past the fiducial marker. Time five oscillations. An oscillation is one complete up and down motion.

Repeat the process for all the different weights.

## The Simulation

Move the lower clamp by placing the mouse pointer over the clamp and using the mouse wheel to move the clamp up and down.

Switch the timer on by mouse clicking the leftmost switch on the timer.

Apply a downward impulse by left-clicking on the weight carrier. You can start timing by clicking on the left button.

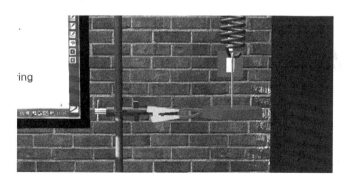

**FIGURE 14.4**
Moving the clamp to indicate the position of the weight cradle.

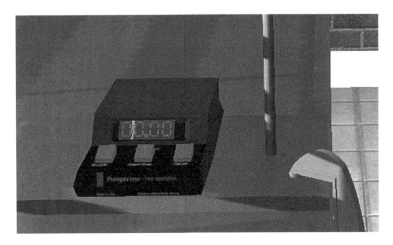

**FIGURE 14.5**
The timer.

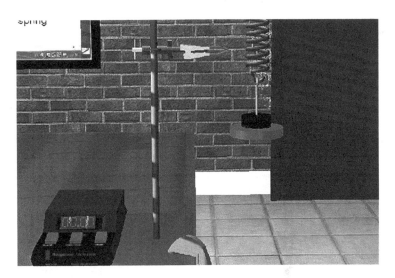

**FIGURE 14.6**
Changing the number of weights on the cradle.

Zooming in using the plus/minus or Z/X keys will make it easier to see the motion.

The carrier without extra weights weighs 50 g. Each additional weight is 50 g. You can add weights by moving the mouse over the carrier and using the mouse wheel.

## The Results

Note the periods for all possible weights and complete the following table:

| Cycles | T /s | P s | P² /s² | M /kg |
|--------|------|------|--------|-------|
| 5 | 5.3 | 1.06 | | |
| 5 | | | | |
| 5 | | | | |
| 5 | | | | |
| 5 | | | | |

When you have completed the table, plot a graph of $T^2$ on the Y-axis against $m$, the mass. This should give you a straight line, which establishes the relationship that the period squared is proportional to the tension in the spring.

Draw the best-fit line through your data points and calculate the gradient.

Use the formula given in the Physics section to calculate the spring constant.

## Further Discussion

One could quite easily find the spring constant by measuring the extensions of the spring under some loads. Do this, then compare the value you obtain to one you calculated using the SHM equation for the period.

Which do you think is more accurate, and why?

# 15

## Capacitor Charge and Discharge

### Introduction

A capacitor is an electronic component that can store a certain amount of charge. It has a value called its capacitance, measured in farads (which is a huge unit; we normally deal with microfarads or even picofarads). The capacitance is the amount of charge a capacitor can store per volt.

- Question: How much charge can a 100µF capacitor store at 6V?

It is an extremely useful device with many applications. For example, it can smooth a power supply's output, or store the charge needed to power a photographic flash (a battery may not be able to deliver the necessary power over the short time period of a flash, but it can trickle charge a capacitor, which can then release the stored charge in one short burst).

This experiment allows you to discover how a capacitor charges and discharges when placed in a circuit. The circuit has a power supply, a resistor, the capacitor, and a two-way switch that can be set to either charge or discharge the capacitor. An ammeter in series with the capacitor monitors the current flowing in or out of the capacitor, and a voltmeter connected in parallel with the capacitor monitors the voltage across it.

### Objective

To establish the relationship between time and the voltage for a capacitor that is charging or discharging.

### The Apparatus

You will need:

- Two multimeters
- A direct current (DC) power supply
- A breadboard

DOI: 10.1201/9781003262350-16

- A Capacitor of 100 microfarads
- A resistor of 100k ohms
- A two- or three-position switch

The circuit, as shown in Figure 15.2, should be constructed on the breadboard.

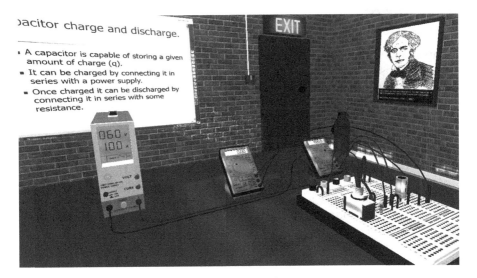

**FIGURE 15.1**
The experiment apparatus.

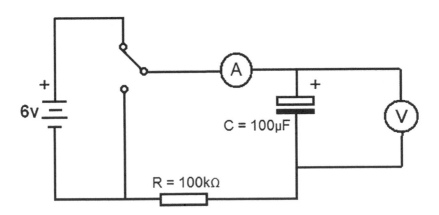

**FIGURE 15.2**
Schematic of the experiment circuit.

## The Circuit

When the switch is in the position shown, the power supply is connected in series with the ammeter and a resistor to the capacitor. A voltmeter is connected across the capacitor. In this configuration, the capacitor will charge. The rate of charge is deliberately slowed by the resistor, which limits the current (by a great deal, having such a high value as 100,000 ohms). This slowing down of the charging process is important, as it allows you to read the voltage at various times in order to plot a graph of voltage against time.

When the switch is in the down position (as indicated on the diagram), the power supply is effectively disconnected from the circuit and the capacitor can now discharge through the ammeter and the resistor.

In both cases, the voltmeter will show the voltage across the capacitor. Note that the voltmeter will affect the circuit very slightly by having its own resistance (which is very high). Also note that the components in the circuit – i.e., the resistor and the capacitor – do not have precise values and there is always some inaccuracy in their values, although they are expected to be within certain tolerances. You can see the tolerances for a resistor quite easily on a resistor color chart that you can look up online. The final band (the lowest, as shown here) indicates the tolerance of the resistor. In the case described next, the band is colored silver.

Question: What are the possible maximum and minimum values for the resistor as given in this circuit, as indicated in the screenshot in Figure 15.3 (the colors from the top are brown, black, yellow, and silver)?

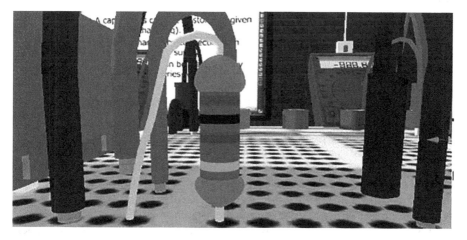

**FIGURE 15.3**
A resistor showing the bands.

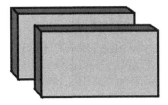

**FIGURE 15.4**
Two parallel conducting plates.

## The Physics

The simplest form of a capacitor is two parallel conducting plates.

When the capacitor is charged, one plate carries a positive charge, and the other plate carries an equal amount of negative charge.

The definition of capacitance is given by the amount of charge on one of the conductors divided by the potential difference between the plates. That is:

$$C = Q/dV$$

Voltage increases linearly with charge, so this quantity is a constant for a given capacitor and is known as the capacitance; it is a measure of the device's ability to store charge.

## Method

With the voltage set to 12V, switch on the power supply.

You need to get ready with a stopwatch at zero; use your wristwatch or your mobile phone, or even the clock from the task bar. With the power on, you now need to click on the switch to start the capacitor charging and start your stopwatch. Take a note of the voltage every 5 seconds for a full minute. After a minute, the capacitor will be fully charged. At this point, you can turn the power supply off.

To discharge the capacitor, click on the two-way switch; you will need to be ready with your stopwatch. Take readings every 5 seconds for a minute.

Note: If you stop charging before the capacitor is fully charged, the capacitor will hold its charge fairly well with just a small amount of leakage, and you can continue charging it at any subsequent time. You will notice that when the power supply is not on but the switch is in the charging position, the voltage across the capacitor will slowly drop. This is because the capacitor is slowly losing charge by leaking through the very high resistance of the voltmeter. There is no current indicated because the ammeter is not part of

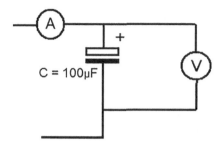

**FIGURE 15.5**
The circuit when the power supply is off, and the capacitor slowly discharges.

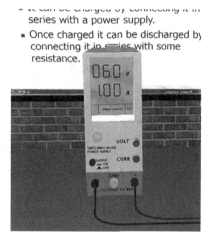

**FIGURE 15.6**
The power supply.

this circuit. Effectively when in this configuration, the circuit is as shown in Figure 15.5.

Only the right-hand part of the circuit is complete, so the current flows. The left-hand part of the circuit is incomplete, so no current flows for the ammeter to detect.

## The Simulation

You can switch on the power supply by clicking the red on/off button. You should see the output indicator on the power supply screen indicate a connection, as shown in Figure 15.6.

If you hover the mouse cursor over the red knob of the switch, you will see its color turn to yellow. This indicates that you can interact with it. When the switch is forward, as shown in Figure 15.7, it is in the discharge position.

The switch is in the discharge position; that is, the two contacts underneath the switch that are closer are connected. Think of the mechanism as that shown in Figure 15.8.

The lever pivots on a hinge and slides a metal contact maker over two of the connecting prongs at a time. When the lever is to the left, the two right prongs are connected, and when it is to the right, the left two prongs are connected.

**FIGURE 15.7**
The switch in the discharge position.

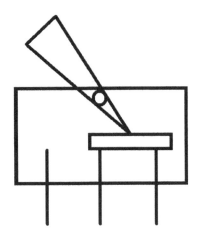

**FIGURE 15.8**
The mechanism of the switch.

## The Results

Plot a graph of voltage on the Y-axis and time on the X-axis for the entire range of timings, both for charging and discharging.

These give the characteristic of an asymptotic curve. On charging, the charging rate slows as it approaches a maximum value. The charging rate is in fact an exponential function. For discharge, the function is:

$$V = V_0 e^{-t/RC}$$

Taking natural logs (logs to the base $e$) of both sides of this equation gives:

$$\ln(V) = \ln(V_0 e^{-t/RC})$$

Multiplication within the brackets of the natural log function is equivalent to addition using the rule:

$$\ln(AB) = \ln(A) + \ln(B)$$

This gives us:

$$\ln(V) = \ln(V_0) + \ln(e^{-t/RC})$$

This simplifies to:

$$\ln(V) = \ln(V_0) - t/RC$$

We can rearrange this to be the equation of a straight line:

$$\ln(V) = -t/RC + \ln(V_0)$$

If you compare this with the standard form of the equation for a straight line:

$$y = mx + c$$

We can verify the discharge equation by plotting $ln(V)$ on the Y-axis and $t$ on the X-axis. The intercept will be $ln(V_0)$ and the gradient will be $1/RC$. Fill in the following table with your results.

| Time /s | I/μA | Voltage/V | ln(V) |
|---------|------|-----------|-------|
| 0 | −51.9 | 5.47 | 1.699279 |
| 5 | −32.6 | 3.44 | 1.235471 |
| 10 | | | |
| 15 | | | |

| Time /s | I/μA | Voltage/V | ln(V) |
|---------|------|-----------|-------|
| 20 | | | |
| 25 | | | |
| 30 | | | |
| 35 | | | |
| 40 | | | |
| 45 | | | |
| 50 | | | |
| 55 | | | |
| 60 | | | |

Plot *ln(V)* on the Y-axis against time on the X-axis. The gradient should be -1/RC. This should give a value for RC somewhere close to 10 (R = 100,000 ohms, C = $10^{-4}$ farads).

## Further Discussion

Find out online how to make a simple capacitor from foil and paper. Try making one and measuring its capacitance; does this conform to what the theoretical calculation of capacitance would give?

Theoretically, the capacitance is given by:

$$C = kA\varepsilon_0/d$$

Where $A$ is the area, $d$ is the separation of the plates, $\varepsilon_0$ is the permittivity of space (8.854 × $10^{-12}$ F/m), and $k$ is the relative permittivity of the dielectric material between the plates. In this case, this is paper, and the value is around 3.8.

# 16

## The Internal Resistance of a Dry Cell

### Introduction

This experiment allows you to measure the internal resistance of a single dry cell using a rheostat, a voltmeter, and an ammeter. The idea of a battery having a resistance can seem counterintuitive; surely the battery is the opposite of a resistor as it is creating a current, not resisting one? However, this is not correct. The chemical processes in a battery create the movement of electrons but these still have to overcome the resistance, no matter how small, of the battery itself. If it helps, think of the battery as something that incorporates a resistor (as shown in Chapter 15); then, you can treat this resistor exactly as you would any other resistor in the circuit.

### The Objective

To find the internal resistance of a dry cell by plotting the relationship between the current and the voltage for a range of rheostat settings.

### The Apparatus

You will need:

- A rheostat
- Two multimeters
- A single dry cell battery and battery box
- Some electrical wire

The device on the left of Figure 16.2 is a rheostat, which is basically a variable resistor that can handle reasonably large currents. The rheostat consists of a large coil of wire (shown in Figure 16.2). Note that there are three connection points: $A$ and $C$ are at each end of the coil, and $B$ is connected by the top metal bar to the sliding connector shown at the middle of the coil. We only need the rheostat connected at $A$ and $B$ to create our variable resistance. When the slider is at the far right, the rheostat is set to its maximum

DOI: 10.1201/9781003262350-17

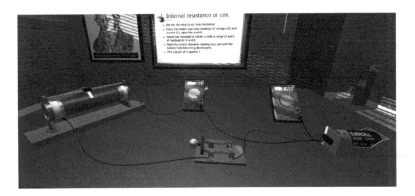

**FIGURE 16.1**
The experiment apparatus.

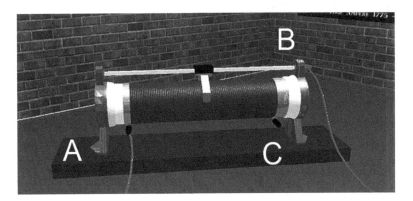

**FIGURE 16.2**
The rheostat.

resistance, as the electricity must go through all the windings of the coil. When the slider is at the left, the rheostat is set to its minimum resistance, as the electricity does not have to go through any of the windings of the coil. You should be careful not to move the slider all the way to the left, as this would cause a short circuit.

## The Circuit

A dry cell is connected in series with a rheostat and an ammeter. A voltmeter is connected across the dry cell (which is also effectively across the rheostat, as the ammeter has negligible resistance).

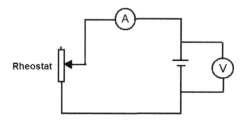

**FIGURE 16.3**
The circuit.

## The Variables

The independent variable is the current that you are controlling with the rheostat. The dependent variable is the voltage.

## The Physics

The energy generated per second within the battery equals the energy transferred into the circuit and into the internal resistance of the battery. Effectively this energy warms the coil of the rheostat and the battery itself, which is why batteries get hot when used (especially high-drain batteries, such as those used in radio-controlled models, which get very hot indeed).

$$\varepsilon I = I^2R + I^2r$$

Where $\varepsilon$ is the electromotive force of the battery, $I$ is the current, $R$ is the resistance of the rheostat, and $r$ is the internal resistance of the cell. Divide both sides through by $I$ to get:

$$\varepsilon = IR + Ir$$

Now rearrange to get:

$$IR = \varepsilon - Ir$$

This is the equation of a straight line with intercept $\varepsilon$ and gradient $-r$. Hence, the gradient of your graph gives you the internal resistance of the cell.

## The Method

Once you have constructed the circuit, set the rheostat to its max resistance by moving the slider as far as you can to the right.

Close the switch and take readings of voltage $V$ and current $I$, then open the switch.

Adjust the rheostat to obtain a wide a range of pairs of readings for $V$ and $I$.

Open the switch between readings (to prevent the battery becoming discharged).

## The Simulation

The first thing you need to do is connect the components together. To do this, you need to click on the start point and end point of each connection in sequence. When you click on where both ends of a wire are connected, the connecting wire will appear. Once all the wires are in place, the circuit will work, and the experiment can begin.

You can move the central contact of the rheostat by positioning the mouse cursor over the middle contact and using the mouse wheel. In the simulation, you cannot move the slider all the way to the left, to prevent the battery short-circuiting.

## The Results

Plot a graph of $V$ against $I$ with $V$ on the Y-axis. The gradient is the internal resistance.

## Further Discussion

Were you surprised at how low the internal resistance is? Can you calculate how much energy (as a percentage) is wasted on heating the battery?

Lithium polymer (LiPo) batteries, used in model cars and planes, get extremely hot while in use. Explain this. Do they have higher-than-normal internal resistance, or is this due to something else?

# 17

# The IV Characteristics of a Diode

## Introduction

This experiment plots the IV characteristics of a diode; that is, you plot how the current changes when you change the voltage (the $I$ is for current, and the $V$ is for voltage). Diodes have the property that they essentially only let current flow in one direction. Most diodes are semiconductor diodes that consist of some semiconductor material with two terminals or connections. Karl Ferdinand Braun discovered the first solid-state semiconductor in 1897 and recognized its use as a rectifier for alternating current (AC). He was also an early pioneer of using these crystals, as they were called, to detect radio waves. The crystal radio sets would have a simple coil and variable condenser for tuning into a particular radio frequency. The signal from this would simply be passed in series through a crystal that would cut out the negative part, allowing the subsequent signal to drive a headphone. Without the crystal diode, the positive and the negative parts of the original signal would cancel each other out. The crystal diode in these early sets would simply consist of the mineral part held rigidly in a holder, and a gold wire, called the cat's whisker, which could be adjusted to make a good contact with the mineral. This is the circuit for a crystal set capable of receiving strong amplitude modulation (AM) signals. Note that the tuning part of the circuit consists of a coil and a variable capacitor connected in parallel. This picks out the

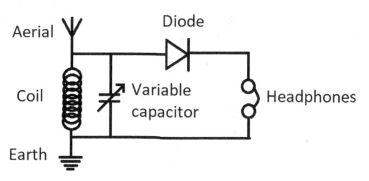

**FIGURE 17.1**
Circuit for a crystal set making use of a diode.

DOI: 10.1201/9781003262350-18

frequency you want to tune to, and then the diode rectifies the signal from that frequency so you can hear it.

The light-emitting variety of diode was first created by Oleg Losev in 1927 and has found many applications, not least of which is low-energy lighting. The reason it is low energy is that we get most of the electrical energy converted to light with very little heat produced.

## Objective

To plot the relationship between the current and the voltage for a diode and establish that a diode only conducts in one direction.

## The Apparatus

You will need:

- A DC power supply
- Two multimeters
- A breadboard;
- An LED
- A 4.7k ohm variable resistor
- A 330 ohm resistor

The circuit, as shown in Figure 17.4, should be constructed on the breadboard.

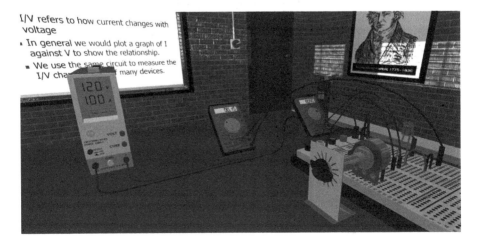

**FIGURE 17.2**
The experiment apparatus.

## The Circuit

The circuit for measuring IV characteristics for any device is mostly the same. A power source is connected in series with a variable resistor, an ammeter, and the device being tested. A voltmeter is connected across the device, which in this case is the LED.

However, for the LED, we are going to modify this circuit to protect the diode from being damaged by too high a current.

Note the 330 ohm resistor in series with the applied voltage and the LED. It should be clear that this circuit is directly measuring the current flowing through the LED and the voltage across it.

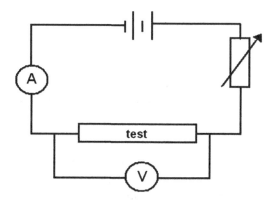

**FIGURE 17.3**
Schematic of the generic IV circuit.

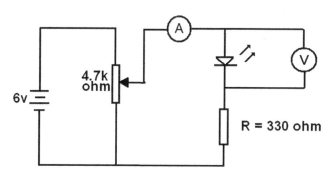

**FIGURE 17.4**
Schematic of the IV circuit used with the diode.

## The Variables

The independent variable is the voltage, and the dependent variable is the current. You are effectively controlling the voltage across the diode.

## Method

Switch on the power. With the power on, you now need to vary the resistance of the variable resistor (sometimes called a potentiometer or even just a pot); this will change the current flowing through the device.

Take readings at every 0.1 volt. Note that the right-hand meter is set to display microamps. To get readings for negative voltages, either reverse the polarity of the power supply or unplug the LED and reconnect it so that its connections are reversed.

## The Simulation

You can switch on the power supply by clicking on the red on/off button. You should see the Output indicator on the power supply screen indicate a connection.

The variable resistor can be varied by placing the mouse cursor over the knob and using the mouse wheel.

To get readings for negative voltage, switch off the power and click on the LED; this will reverse its connections in the circuit. Now turn the power on and repeat the readings taken previously. The voltages this time count as negative, as the LED is now connected the other way around.

## The Results

Plot a graph of current in amps on the Y-axis and volts on the X-axis for the entire range of voltage readings. You should see an L-shaped plot with the current zero through all negative values until the diode switches on at some positive value.

## Further Discussion

Find a specification sheet online for the LED you are using. How closely does your graph compare to the specification?

# 18

## The IV Characteristics of a Filament Lightbulb

### Introduction

This experiment plots the IV characteristics of a lightbulb; that is, you can plot how the current changes when you change the voltage (the *I* is for current and the *V* is for voltage). This depends on the resistance of the bulb, which changes depending on the voltage. Effectively, the bulb gets hotter and hotter as the voltage increases, until it is eventually white-hot. As the temperature of the metal filament gets hotter, its resistance changes. As long ago as 1802, Humphry Davy demonstrated a filament light at the Royal Institution. However, it is usually Thomas Edison who gets the credit much later (he wasn't born until 1847!), as he was able to come up with an entire practical system for electric lighting. Edison was a prolific inventor and held over one thousand patents.

### The Objective

To plot the relationship between the current and the voltage for a filament lightbulb.

### The Apparatus

You will need:

- A DC power supply
- Two multimeters
- A breadboard
- A 12V filament lightbulb
- A 4.7k ohm variable resistor

The circuit, as shown in Figure 18.2, should be constructed on the breadboard.

DOI: 10.1201/9781003262350-19

**FIGURE 18.1**
The experiment apparatus.

---

## The Variables

The independent variable is the current, and the dependent variable is the voltage.

---

## The Circuit

The circuit for measuring IV characteristics for any device is nearly always the same. A power source is connected in series with a variable resistor, an ammeter, and the device being tested. A voltmeter is connected across the device.

---

## The Method

Switch on the power supply. With the power on, you now need to vary the resistance of the variable resistor (sometimes called a potentiometer, or even just a pot); this will change the current flowing through the device.

Take readings of the current at every 0.5 volt up to 12V.

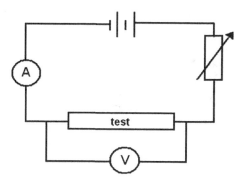

**FIGURE 18.2**
The experiment circuit.

**FIGURE 18.3**
The power supply.

## The Simulation

You can switch on the power supply by clicking on the red on/off button. You should see the Output indicator on the power supply screen indicate a connection, as shown in Figure 18.3.

The variable resistor can be varied by placing the mouse cursor over the knob and using the mouse wheel.

Note that the current is displayed on the right-hand meter and is displayed in milliamps.

**FIGURE 18.4**
The variable resistor.

## The Results

Plot a graph of current in amps on the Y-axis and volts on the X-axis. What you should see is the voltage and amperage both increasing together, with the current tailing off as the filament gets hotter and more resistant.

## Further Discussion

What is the physical reason that explains why the resistance of the metal filament increases with temperature?

# 19

## The Resistivity of Constantan

DOI: 10.1201/9781003262350-20

### Introduction

Every material that obeys Ohm's law (not all materials do) has a characteristic resistivity. The resistivity is a constant for a particular substance that allows you to calculate what the resistance is in ohms for a wire of a particular length with a particular cross-sectional area. This is given by:

$$R = \rho L/A$$

Where $\rho$ is the resistivity, $L$ is the length, and $A$ is the cross-sectional area.

This experiment allows you to plot the resistance against the length of wire that the current is flowing through. Given that you can find the cross-sectional area by measuring the diameter with a micrometer, you can then calculate the resistivity.

### Objective

To find the resistivity of constantan.

### The Apparatus

You will need:

- A DC power supply
- A length of constantan wire
- Two multimeters
- A micrometer
- A meter ruler
- Two crocodile clips
- Some ordinary electrical wire
- Two mounting blocks to support the wire

The circuit should be wired as shown in Figure 19.2.

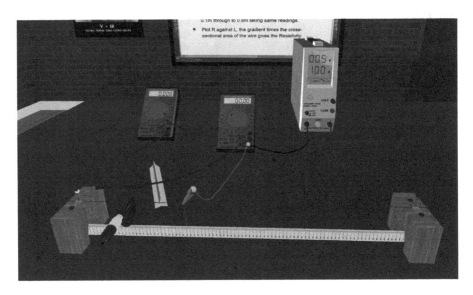

**FIGURE 19.1**
The experiment apparatus.

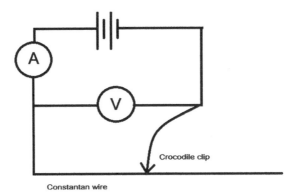

**FIGURE 19.2**
The experiment circuit.

## The Circuit

A power source is connected in series with the constantan wire and the ammeter. A voltmeter is connected across the wire. A crocodile clip is used so that the length of the wire that forms part of the completed circuit can be varied.

## The Variables

The length of the wire in the closed circuit is the independent variable. The voltage and the current are the dependent variables.

## The Physics

As given in the introduction to this chapter, the resistance of a length of wire is given by:

$$R = \rho L/A$$

Where $\rho$ is the resistivity, $L$ is the length, and $A$ is the cross-sectional area. This can be rearranged to give an equation for the resistivity:

$$\rho = RA/L$$

Given that we can calculate $A$ from the diameter of the wire, plotting $R$ against $L$ will give us a gradient that we can use to calculate the resistivity.

## The Method

Use the micrometer to measure the diameter of the constantan wire. Make a note of this in your laboratory book.

Switch on the power supply. With the power on, you now need to vary the voltage to 0.5V. Move the crocodile to the 10-cm point.

At this point, take readings of the voltage and current. Now move the crocodile clip to the 20-cm position and increase the voltage by 0.5V (this is to keep the current roughly the same for each reading). Take readings for voltage and current. Repeat this for lengths of constantan up to 80 cm.

## The Simulation

You can switch on the power supply by clicking on the red on/off button. You should see the Output indicator on the power supply screen indicate a connection, as shown in Figure 19.3.

If you hover the mouse over the red knob marked 'VOLT', you will see it change to yellow. Using the mouse wheel, you can raise or lower the voltage output by the power supply unit.

You can move the position of the crocodile clip that is attached to the wire by placing the cursor over the clip and using the mouse wheel.

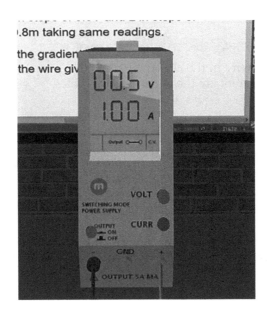

**FIGURE 19.3**
The power supply.

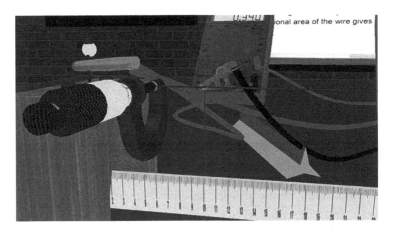

**FIGURE 19.4**
Using the micrometer to measure the diameter of the wire.

At the start of the experiment, you will need to move the crocodile to the 10-cm point, so you will first need to move the micrometer out of the way. With the mouse over the jaws of the micrometer, rotate the mouse wheel so that the micrometer moves to the left; move it as far as you can. Now, using the same method, move the crocodile clip to the 10-cm point. Notice that there is an indicator which shows its position on the ruler underneath, as shown in Figure 19.4.

You will need to measure the diameter of the wire using the micrometer. To do this, move the crocodile clip as far to the right as you can so that it is well out of the way. Then position yourself at the left-hand edge of the table. Use the zoom function (plus and minus or Z/X keys) to zoom in on the micrometer. You can rotate the barrel using the mouse wheel when the cursor is over the end part of the barrel. The rachet will sound when the micrometer is tight. Tighten the micrometer onto the wire by rotating the barrel, and then take the reading of the diameter. You cannot move the micrometer when it is tightly gripping the wire.

Do this in two other places along the wire and take the average.

## The Results

Complete this table with your readings:

| L /m | V /volts | A /Amps | R /Ohms |
|------|----------|---------|---------|
|      |          |         |         |
| 0.1  | 0.32     | 0.34    | 0.941176 |
| 0.2  |          |         |         |
| 0.3  |          |         |         |
| 0.4  |          |         |         |
| 0.5  |          |         |         |
| 0.6  |          |         |         |
| 0.7  |          |         |         |
| 0.8  |          |         |         |

You can calculate the ohms column using Ohm's law. A trick for remembering Ohm's law is to think of the name Eric. $I_c$ is often used to denote current, $E$ for electromotive force, and $r$ for resistance. So the formula can be remembered as:

$$E = R\,I_c$$

When you have completed the table, plot a graph of $R$ against $L$ and determine the best-fit line through your data points.

Where $R/L$ is the gradient of your graph, the resistivity is calculated from this gradient multiplied by the cross-sectional area.

## Further Discussion

Why do we try to keep the current the same for each reading that we take?

# 20

## Resistors in Series and Parallel

DOI: 10.1201/9781003262350-21

### Introduction

Resistors impede the flow of current in a circuit. We use them in electrical and electronic circuits to control the flow of current. The higher the resistance, the less current that flows. However, there are two different ways we can connect resistors together, either in series or in parallel. This experiment will allow you to investigate the combined resistance of resistors connected both ways.

Resistance is measured in ohms, the symbol for which is the Greek letter omega, $\Omega$. The unit is named after Georg Simon Ohm, who was a German physicist and mathematician. He discovered that there is a direct relationship between potential difference across a conductor and the amount of current that flows. This is known as Ohm's law.

### Objective

To investigate the combined resistance of resistors in series and in parallel and to confirm the rules for calculating the combined resistance of resistors in series and parallel.

### The Apparatus

You will need:

- A breadboard
- A battery
- Some resistors of the same and different values
- Some electrical wire
- A multimeter or an ammeter

The breadboard needs to be wired as shown in Figure 20.4 and connected to the battery and ammeter as shown in Figure 20.5.

## The Variables

The current is the dependent variable, and the independent variables are the resistors used in the circuit.

## The Physics

Ohm's law gives us the means of calculating the voltage given the current and resistance in a simple circuit.

The law is:

$$E = IR$$

Where $E$ is the electromotive force (potential difference or voltage) in volts, $I$ is the current in amps, and $R$ is the resistance in ohms. The equation can be rearranged to give us any one of the three factors, given the other two.

Resistors in series like that shown in Figure 20.2 have a combined resistance of the sum of their individual resistances. In this case:

$$R = R1 + R2$$

The combined value can then be used in Ohm's law to calculate either the voltage or the current, given that one of them is known. But note that if the voltage and current are known, as would be the case in the previously described circuit, then only the combined resistance can be calculated.

For resistors in parallel, as shown in Figure 20.3, the combined resistance is given by the equation:

$$1/R = 1/R1 + 1/R2 + \ldots$$

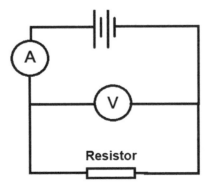

**FIGURE 20.1**
A simple resistor circuit to measure the current and voltage.

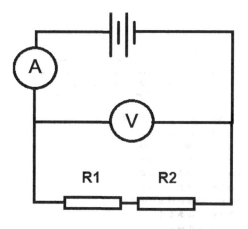

**FIGURE 20.2**
Resistors in series.

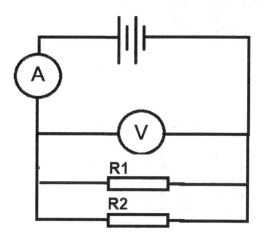

**FIGURE 20.3**
Resistors in parallel.

These two laws for calculating the total resistance of resistors in series and in parallel are what you are going to verify in this experiment.

## The Method

The breadboard should be initially wired as shown in Figure 20.4.

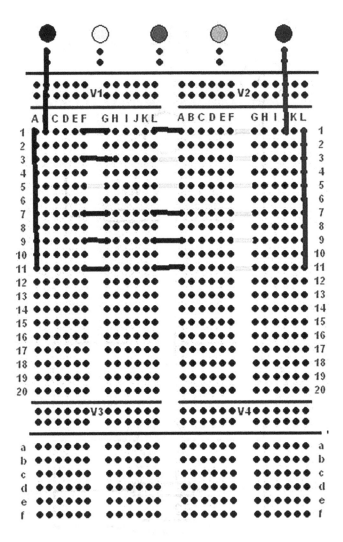

**FIGURE 20.4**
The breadboard wiring.

This configuration allows for row 1 to be used for a single resistor, row 3 for two resistors in series, row 5 for three resistors in series, row 7 for a second resistor in parallel (with row 1), row 9 for three resistors in parallel, and row 11 for four resistors in parallel.

The breadboard contacts should be wired in series with the ammeter (or multimeter measuring current) and the battery.

Resistors should now be placed into the breadboard to give configurations for resistors in series and parallel.

For the initial investigation, you should use resistors of the same value. There are three 33 ohm resistors at the nearest end of the cork mat; they are colored orange, orange, black (orange is for 3 and black is for 1, which is the multiplier). This is how we identify the value of a resistor. Resistors have preferred values, so only some combinations are used. You can look up a resistor color code chart online or use the one inside the simulation.

## The Simulation

Click on a resistor and then click on the breadboard for where you want the resistor to be placed.

You can return a resistor to the cork mat by clicking on it while it is positioned on the breadboard.

The multimeter will read the current for the circuit, whatever the configuration.

Using only the 33 ohm resistors, place one on row 1 where the yellow strip is. Record the current and calculate the resistance of the resistor using the assumed voltage of the battery and the current reading. Return the resistor to the cork mat. Now place two of the 33 ohm resistors in series on row 3 of the breadboard, record the current and calculate the combined resistance from it. Return the two resistors to the cork mat. Now place three of the 33 ohm resistors onto row 5 of the breadboard. Again, calculate their combined resistance from the recorded current.

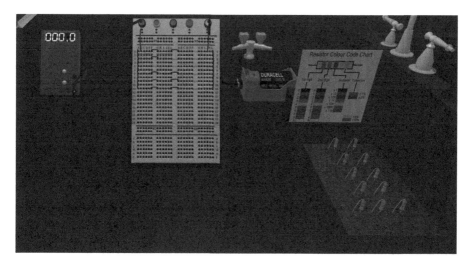

**FIGURE 20.5**
View of the simulation showing the breadboard, multimeter, and battery.

## The Results

For resistors connected in series, you should be able to verify that the total resistance in each case is the sum of the individual resistances.

For resistors connected in parallel, you need to confirm the more complicated formula.

$$1/R = 1/R1 + 1/R2 + \ldots$$

## Further Discussion

Can you confirm the two laws using a mixture of resistance values?

What happens when you mix resistors in parallel with resistors in series?

# 21

## Heat Transfer

### Introduction

This simple experiment allows you to determine how different materials affect heat loss. We are all used to the idea of having heating on indoors during the winter when the outside temperature is low. We commonly use double glazing to make the insulation between the inside and outside of our houses stop as much heat loss as possible. All substances conduct heat at different rates; some are much better insulators than others. In Figure 21.1, we can see that the temperature drops from an inside warmer temperature to a colder temperature across the insulating substance.

### The Objective

To investigate the insulation properties of different substances.

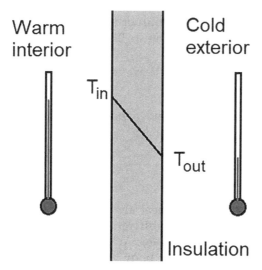

**FIGURE 21.1**
The temperature gradient through an insulating layer between a warm interior and a cold exterior.

DOI: 10.1201/9781003262350-22

**FIGURE 21.2**
Lighting the Bunsen burner.

## The Apparatus

You will need:

- A Bunsen burner, gauze, and stand
- A thermometer
- A beaker of water with a thick cork lid
- A set of 'wraps' made of different material that can be placed over the beaker

Put the stand over the Bunsen burner and place the gauze and beaker of water on the stand as shown in Figure 21.2.

## The Variables

The insulation material is the independent variable, and the temperature and time are the dependent variables.

## The Physics

Thermodynamics tells us that heat always flows from a hotter area to an adjacent cooler area. In general, different substances will allow heat to conduct through them at different rates. When we heat one side of a body, the energy will transfer through the body to the other side and then dissipate. There will be a temperature gradient through the body.

Where a fire inside a house is heating a window, the temperature on the inside of the window is relatively high when compared to the temperature on the outside of the window. Throughout the wall is a temperature gradient joining the inside temperature to the outside temperature. Heat is dissipated into the air on the outside.

The ease with which heat can transfer through a body is called its thermal conductivity.

## The Method

Heat the water in the beaker to boiling or close to the boiling point. Turn off the Bunsen burner and wait for 1 minute, and then record the temperature. Wait for 5 minutes and record the temperature again. Now repeat the experiment for the other remaining materials.

## The Simulation

Light the Bunsen burner by clicking on the gas tap (Figure 21.2).

The materials are cork, newspaper, and wool. A material wrap can be used or removed by clicking on it. Heat the water until it nearly boils, then switch off the gas and let the water cool for a minute. Take the temperature. Now wait an additional 5 minutes and take the temperature again.

Repeat for the other three material wraps.

## The Results

Complete this table with your readings:

|  | T°C At 60 Secs | T°C At 360 Secs |
|---|---|---|
| No Insulation |  |  |
| Cork |  |  |
| Newspaper |  |  |
| Wool |  |  |

From the table you can identify the best and worst insulators.

## Further Discussion

Without any insulation, heat the water to as hot as you can get it. Then turn off the Bunsen burner and record the temperature in an Excel spreadsheet every 20 seconds for a full 10 minutes. Your spreadsheet will look something like Figure 21.3, although it will be longer.

| ◢ | A | B | C | D |
|---|---|---|---|---|
| 15 | | | | |
| 16 | | | | |
| 17 | | Newton's law of cooling | | |
| 18 | | | | |
| 19 | | 0 | 99.2 | |
| 20 | | 20 | 96.1 | |
| 21 | | 40 | 93.4 | |
| 22 | | 60 | 90.9 | |
| 23 | | 80 | 88.3 | |
| 24 | | 100 | 85.8 | |

**FIGURE 21.3**
The results of cooling a liquid in a spreadsheet.

When you have all the data, plot it using the Insert/Scatter graph function in Excel.

What do you see?

Is the cooling completely linear, or does the cooling rate slow down?

Can you explain this?

# 22

## Boyle's Law

### Introduction

Robert Boyle was born in Ireland and lived during the seventeenth century; he had considerable wealth, which allowed him to spend a great deal of it investigating various physical phenomena. He published the second edition of *New Experiments Physico-Mechanicall, Touching the Spring of the Air and its Effects* in 1662, which contained a description of the experiment that showed what we now call Boyle's law.

Boyle's law simply states that for a gas at constant temperature, the product of pressure and volume is a constant. That is:

$$PV = \text{constant}$$

If we take several readings at different volumes, then we expect the result:

$$P_1V_1 = P_2V_2$$

This experiment will verify this equation. Boyle's apparatus consisted of a U-tube sealed at one end, into which he poured mercury and examined the volume of the trapped air as the amount of mercury was increased. Mercury, being both very heavy and liquid at room temperature, was ideal for the job.

The pressure from the weight of the mercury increased as the volume of air decreased. In Figure 22.1, the gas in the sealed end must be at atmospheric pressure as the level of the mercury is level. Think of it this way: if the forces acting on both ends of the mercury were not equal, then the mercury would move. As mercury is added, the gas in the sealed-off end becomes compressed.

The pressure on the gas in the sealed end is now equal to the atmospheric pressure plus the pressure from the weight of the mercury that is above the level on the left of the tube.

Robert Hooke was working for Boyle when this experiment was performed, and it was Hooke's careful measurements that resulted in a table of observations that showed the relationship now known as Boyle's law. Although not used here, it was Hooke who designed and made the vacuum pumps that Boyle used in his other experiments with gases and the atmosphere.

DOI: 10.1201/9781003262350-23

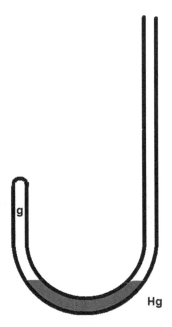

**FIGURE 22.1**
Mercury in a U-tube sealed at one end.

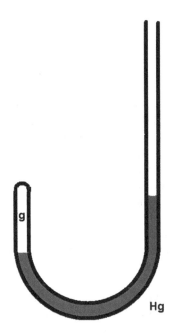

**FIGURE 22.2**
Additional mercury in a U-tube so it is no longer level.

## The Objective

To verify Boyle's law: that at constant temperature, the product of pressure and volume is a constant.

## The Apparatus

You will need:

- A 10-ml syringe
- A stand with adjustable clamp
- Weight cradle and set of weights
- A G-clamp
- A micrometer
- Some rubber tubing sealed off at one end

Assemble the apparatus as shown in Figure 22.3.

**FIGURE 22.3**
The experiment apparatus.

A syringe is held by a clamp, with a weight holder attached to the plunger so that a force can be applied to it changing the pressure. Notice that the syringe is closed off at the top with a length of rubber tube and a small clamp. The stand can be clamped to the table with the G-clamp to give it added stability. A ruler is placed behind the syringe so that the position of the plunger can be easily determined.

## The Variables

The pressure is the independent variable, and the volume is the dependent variable.

## The Physics

The area of the plunger can be calculated from its radius using the formula:

$$A = \pi R^2$$

The pressure can be calculated from the formula:

$$P = F/A$$

Where $F$, the force, is calculated from the mass times $g$, the acceleration due to gravity (9.81 m/s²).

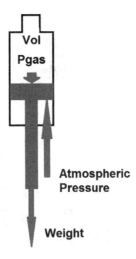

**FIGURE 22.4**
The forces on the plunger.

To calculate the pressure in the syringe, you need to take account of the standard atmospheric pressure of 101 kPa. The pressure of the gas inside the syringe is the difference between the standard atmospheric pressure and the pressure due to the weight, as calculated previously.

## The Method

Accurately measure the diameter of the plunger head with a micrometer. Make a note of your measurement. Seal off the end of the syringe with a short length of tubing and a small clamp. Alternatively, seal it off with some hard-setting glue. For weights of 200 g, 400 g, 600 g, 800 g, and 1000 g, read the volume as accurately as you can from the position of the plunger in the syringe.

## The Simulation

Measure the diameter of the plunger head by using the mouse wheel with the cursor over the barrel of the micrometer. Turn the barrel with the wheel until it will not turn any further. At this point, the micrometer is gripping the plunger. Read the micrometer to get the diameter.

The weights can be altered by putting the mouse cursor over the weight cradle and using the mouse wheel.

You can zoom in and out using the plus and minus or Z and X keys. Measure the plunger position as shown in Figure 22.5.

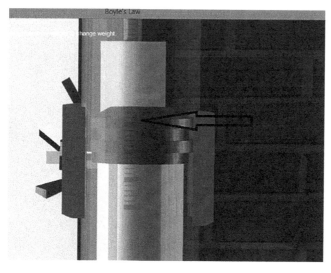

**FIGURE 22.5**
Observing the plunger's position.

## The Results

Create a table as follows but with your own data:

| Mass /kg | Volume /ml | P mass /Pa | 1/V m⁻³ | P syringe /Pa |
|---|---|---|---|---|
| 0.2 | 4.38 | 8.45E+03 | 228310.5 | 9.26E+04 |
| 0.4 | | | | |
| 0.6 | | | | |
| 0.8 | | | | |
| 1 | | | | |

When you have completed the table, plot 1/V against the pressure. You should obtain a straight line, which confirms the law.

## Further Discussion

How do you think this experiment might be improved?

In particular, how do you think it could be made more accurate?

What do you think of Boyle's original method?

# 23

## Charles's Law

### Introduction

Charles's law states that at constant pressure, the volume of a fixed amount of a gas is proportional to the temperature.

The law takes its name from Jacques Charles, who discovered the law in the 1780s but did not publish the result. Joseph Gay-Lussac discovered the law independently. In 1801, John Dalton demonstrated that the law applied to all gases.

### The Objective

The objective is to confirm Charles's law that the volume of a fixed amount of gas increases in direct proportion to the temperature, or:

$$V = kT$$

Additionally, to estimate a value of absolute zero in degrees centigrade.

### The Apparatus

You will need:

- A glass beaker half filled with boiling water
- A glass capillary tube closed at the bottom end, containing a drop of oil
- A digital thermometer
- A ruler
- A tap that can be used to add cold water

The ruler, thermometer, and capilliary tube should be placed in the beaker of boiling water.

DOI: 10.1201/9781003262350-24

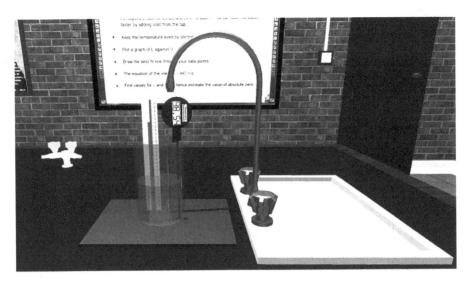

**FIGURE 23.1**
The apparatus for Charles's experiment.

## The Variables

The independent variable is the temperature, and the dependent variable is the volume of the air.

## The Physics

Charles's law relates temperature to volume of a gas at constant pressure. Having the capillary tube open at the top end ensures that the trapped air in the tube is kept at constant pressure, which is the current atmospheric pressure within the laboratory.

Over time, the water in the beaker will cool, and because the part of the capillary tube containing the trapped air is totally submerged in the water, it will also be cooled to the same temperature as the water.

## The Method

At five-degree intervals, take the temperature of the water and the length of the trapped volume of air. As the temperature of the water approaches that

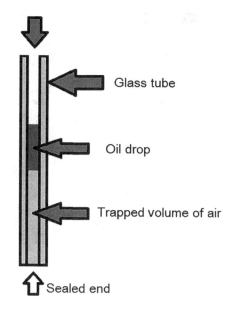

**FIGURE 23.2**
Close-up of the capillary tube.

of the laboratory, the cooling rate will slow. You can cool the water more rapidly by adding cold water from a tap.

If you add cold water, you will need to ensure that the water in the beaker is well mixed; otherwise, you will get erroneous readings for the temperature. You can stir the water using the thermometer.

## The Simulation

If you place the mouse cursor over the nearest tap, you can use the mouse wheel to add cold water to the beaker.

To stir the water with the thermometer, put the mouse cursor over the thermometer and use a forward and backward motion with the mouse wheel to stir the water.

To get an accurate reading of the length of the air column, move in as close as you can and use the page up/down keys to get on a level with the oil drop, and use the magnify view buttons (numeric +/- or Z/X) to zoom in.

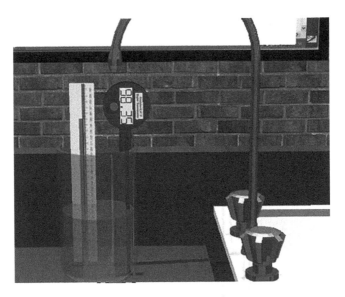

**FIGURE 23.3**
Adding water to the beaker.

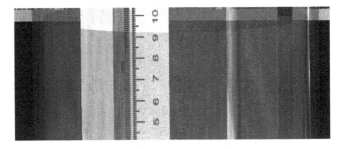

**FIGURE 23.4**
Close-up of measuring the length of the air column.

## The Results

Use the readings to complete this table:

| T /° C | L /cm |
|--------|-------|
| 95     | 7.28  |
| 90     |       |
| 85     |       |
| 80     |       |
| 75     |       |

| T /° C | L /cm |
|--------|-------|
| 70 | |
| 65 | |
| 60 | |
| 55 | |
| 50 | |
| 45 | |
| 40 | |
| 35 | |
| 30 | |
| 25 | |

You can add a column for volume and calculate the volumes from the lengths. When you have completed the table, plot a graph of *V* against *T*. Draw the best-fit straight line through your data points. The straight line vindicates Charles's law.

## Further Discussion

We can go on to estimate a value for absolute zero in centigrade.

From your graph, calculate the gradient of the line. The equation relating *V* against *T* is:

$$V = mT + c$$

Having calculated the gradient, *m*, from your graph, you can now calculate *c* by substituting a value pair for *V* and *T* into the equation.

Now by substituting a value of zero for *V* and solving the equation for *T*, you can find a value for absolute zero.

# 24

## Mechanical Equivalent of Heat

### Introduction

It may seem obvious that heat energy and mechanical energy are the same thing. However, it wasn't always so. In the middle of the nineteenth century, James Prescott Joule's experiment that measured this equivalence was groundbreaking and used state-of-the-art thermometers to measure the tiny differences in temperature obtained. Indeed, it is not entirely clear how Joule made such accurate measurements. His thermometer (built by Henri Victor Regnault) had ten divisions per degree Fahrenheit, and these ten divisions measured 0.5 inches (about 1.27 cm). So one division was equivalent to 0.1 °F. Joule claimed that he could quite easily measure to 1 twentieth of a division, allowing him to measure to 1 two-hundredth of a degree.

At this time, the nature of heat was not fully understood, and heat was commonly attributed to a substance called caloric that could be taken away

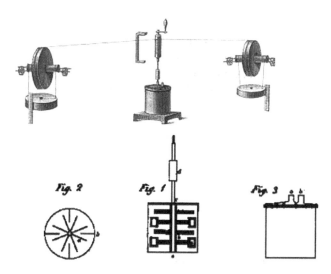

**FIGURE 24.1**
Joule's original apparatus.

DOI: 10.1201/9781003262350-25

or added to anything that could be heated or cooled. For example, caloric plus ice made water.

Joule's original apparatus used weights that fell, driving a cord around an axle that, in turn, rotated a set of paddles inside a container of water.

There was a handle enabling him to wind the weights back to the top so that the mechanical energy could be repeatedly applied.

The weight and the distance fallen allowed him to calculate the mechanical energy. Careful measurement of the temperature of the water and the volume of the water allowed him to calculate the energy gained.

Our apparatus is rather simpler. We generate the heat by the friction between a cord and a brass cylinder, as described in the next section.

## The Apparatus

You will need:

- A brass cylinder with a handle to rotate it and a counter to count the number of rotations
- A rope that is fixed at one end to the apparatus, is wrapped around the cylinder several times, and is attached to a weight that holds it taut
- A digital thermometer that is embedded into the brass cylinder
- A room thermometer that supplies the ambient temperature
- A G-clamp

The apparatus that contains the counter, drum, and handle should be firmly fixed to the workbench using a G-clamp. If you cannot source the complete cylinder and counter unit, this can be made from individual components: a brass drum drilled to take a shaft and handle with a hole for a thermometer and a mechanical revolution counter, all held in place with a suitable bracket.

The rope should be wrapped around the brass cylinder eight to twelve times. One end should be fixed, as shown in Figure 24.2. The other end should be attached to a weight and allowed to dangle, as shown in Figure 24.3.

## The Variables

The independent variable is the number of turns made of the drum. The dependent variable is the temperature rise of the drum.

**FIGURE 24.2**
The counter, drum, and handle.

**FIGURE 24.3**
The rope and weight.

## The Physics

The work done, *W*, is given by force times distance:

$$W = MgRN2\pi$$

Where *M* is the mass of the weight, *g* is the acceleration due to gravity, *R* is the radius of the cylinder, and *N* is the number of turns. The result is in joules.

The heat gained, *Q*, is given by:

$$Q = M_c S_b \Delta T$$

Where $M_c$ is the mass of the cylinder, $S_b$ is the specific heat capacity of the cylinder, and $\Delta T$ is the change in temperature. The result is in calories.

Now, to calculate the work in joules needed to supply one calorie of heat, we need to calculate the ratio:

$$W/Q$$

This is the result.

## The Method

When the experiment is started, the room temperature should be a little above that of the cylinder and the cylinder will start to warm very slowly. When the cylinder is above room temperature, it will cool very slowly. In order to minimize the effects of this possible gain or loss of heat to the room, it is recommended to take the cylinder to a temperature that is the same amount above room temperature as it starts off below. In this way, the amount of heat gained by the cylinder from the room will be the same as that lost to the room.

Turn the handle (this will probably take a few hundred turns) until the final temperature is the same amount above room temperature as it was below it at the start. Note these temperatures and the number of turns needed. You can calculate $\Delta T$ from the difference in temperatures at the start and end of the turning. Note, take these readings immediately before you start turning and immediately at the end before the cylinder starts to lose heat to the room.

## The Simulation

Notice that there is a thermometer in the room giving the room temperature.

**FIGURE 24.4**
The room thermometer.

**FIGURE 24.5**
The cylinder thermometer.

The handle can be turned by placing the mouse over the area taken by the handle. You need to position yourself so that you can see the readings on the thermometer. Note the number of turns counter on the top of the apparatus.

## The Results

The result is simply the ratio of W/Q that we showed how to calculate in the Physics section.

## Further Discussion

An actual experiment would show a slight rise in temperature after the turning stops due to heat conducting from the surface throughout the brass cylinder. This effect has not been included in the simulation. However, if one is do an actual experiment, it is worth taking note of.

Comment on any further heat losses that may affect the experimental results during this period.

Are there any other methods you might propose by which this experiment could be improved?

How do you think Joule could have made his observations that he claimed? He wrote:

> When this was done it was found that the ten divisions of the sensible thermometer (occupying about half an inch) were nearly equal to the degree of Fahrenheit . . . since constant practice has enabled me to read off with naked eye to 1/20th of a division.[1]

## Note

1. Joule, James Prescott. "On the Mechanical Equivalent of Heat." *Philosophical Transactions of the Royal Society of London* (London: The Royal Society, 1850), vol. 140, pp. 61–82, http://www.jstor.org/stable/108427.

# 25

## Specific Heat Capacity of Brass

## Introduction

The specific heat of a substance is the amount of heat it can hold per unit mass. In this experiment, we place a brass weight in boiling water. When the weight is in the boiling water, it is gaining heat energy. After about 4 or 5 minutes, it will be heated evenly to the temperature of the water. It is at this point that it has gained all the heat that it can from the boiling water.

The brass weight can then be moved to the water in the calorimeter. It will then heat the water and the calorimeter.

Knowing the specific heat capacity of the water and calorimeter and the maximum temperature attained allows you to calculate the heat gained by them. This amount of heat gained must be the same as the amount of heat lost by the brass. This enables the specific heat capacity of brass to be calculated.

## The Objective

The aim is to find $Sb$, the specific heat of brass.

## The Apparatus

You will need:

- A glass beaker three-quarters filled with water
- A brass weight (100 g)
- A Bunsen burner, tripod, and gauze to keep the water boiling
- A calorimeter made of copper with a known quantity of water
- A digital thermometer
- An orbital shaker

Configure the apparatus as shown in Figure 25.1. The orbital shaker is optional, as it is sufficient to stir the water.

DOI: 10.1201/9781003262350-26

**FIGURE 25.1**
The experiment apparatus.

## The Variables

The time is the independent variable, and the temperature is the dependent variable.

## The Physics

The heat gained by the water and the calorimeter equals the heat lost by the brass weight.

If the mass of the calorimeter is $M_c$, the specific heat capacity of brass is $S_b$, the mass of the water is $M_w$, the specific heat of water is $S_w$, the mass of the brass weight is $M_b$, the initial temperature of the water is $T_i$, and the final temperature of the water is $T_f$, then:

$$M_cS_b(T_f - T_i) + M_wS_w(T_f - T_i) = M_bS_b(100 - T_f)$$

## The Method

Bring a beaker of water to boiling using a Bunsen burner. Use a loop of wire or string to lower the brass weight into the water.

Heat the brass weight to 100 °C by leaving it in the boiling water for 5 minutes. Then plunge it into the cold water in the calorimeter. Start the shaker to even out the heating of the water.

Note the temperature of the water before adding the weight and then at 15-second intervals.

Determine the highest temperature and therefore the heat gained by the water and calorimeter and the heat loss from the brass. A correction can be made to the directly observed maximum by noting that there is an inevitable heat loss from the calorimeter to the room. If we record the temperature for several minutes after the maximum has been reached, we can tell how much heat is being lost to the room per minute and use that to correct the observed maximum.

## The Simulation

You can lift the brass weight by placing the mouse cursor over the card attached to the weight and left-clicking. You can then move the weight using the mouse. If the weight collides with anything, it will be released immediately. You can also release by clicking again. The distance from you can be changed by holding down the shift key and using the mouse scroll wheel. Note that the weight's shadow on the desk will help you to determine its position above the desk. In particular, you should look for the shadow on the rim of the calorimeter, which you can see clearly in Figure 25.2.

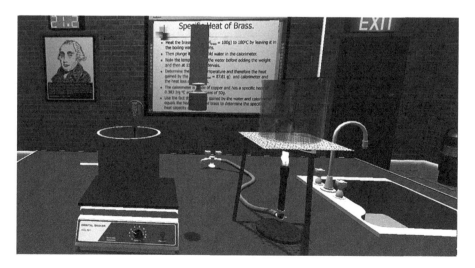

**FIGURE 25.2**
The shadow cast by the brass weight.

## The Results

The calorimeter is made of copper, so you can calculate the heat gained by it.

Use the fact that the heat gained by the water and calorimeter equals the heat lost by the brass to determine the specific heat capacity of brass.

## Further Discussion

In an actual experiment, we would try to improve the accuracy by tapping the brass on the bench before plunging it into the water in the calorimeter to remove any hot water that might cling to it.

We would also probably stir the water in the calorimeter to ensure that the heating was even rather than use an electrical shaker.

Can you find any other suggestions for improving the accuracy of this experiment?

# 26

## Investigation of Mechanical Waves

### Introduction

A wave is an oscillation of something that transfers energy along its path of propagation without transferring any matter. Think of a wave produced by flicking the end of a long rope. A wave will travel along the rope, but no part of the rope is actually moving in the direction of the wave; the parts of the rope are simply going up and down.

If we tied the far end of the rope to a fixed point and oscillated the near end, we would set up a pattern of moving waves on the rope. This is a transverse wave, where the direction of the wave motion is at right angles to the disturbance, which is up and down. This is also like the wave motion in the sea, where the water is going up and down and a surfer can ride the crest of a wave and be propelled forward in the direction of propagation of the wave. The distance between successive crests is called the wavelength. The number of complete waves per second is called the frequency, measured in hertz (Hz). One Hz is one cycle (complete wave) per second. The wavelength is often represented by the Greek letter lambda ($\lambda$) and the frequency by $v$. There is a simple relationship between the speed of the wave, its frequency, and its wavelength:

$$v = f\lambda$$

This is easiest to understand if you just consider the case where the frequency is one cycle per second. In this case, the wave must propagate exactly one wavelength each second.

The kind of waves we are considering here are called mechanical waves, as something physical is moving. This is different from electromagnetic waves, which can travel through a vacuum and don't involve the motion of anything physical. In our experiment, we are going to use an apparatus first proposed by John Shive of Bell Laboratories in 1959, which he called the mechanical wave machine. This consisted of wires along which the waves travel, not as a vertical displacement but rather as a twisting of the wire. Along the wire are placed horizontal pendulums; these have the effect of slowing the wave motion, allowing the wave propagation to be seen more easily.

DOI: 10.1201/9781003262350-27

We will use a particularly simple version of this where we use a central tape instead of a wire, which is attached to fifty horizontal pendulums. Note that in this apparatus, it is the twisting of the tape that propagates the wave.

## The Objective

The aim is to find the relationship between the tension and the speed of propagation of the wave.

## The Apparatus

You will need:

- The mechanical wave machine
- A frequency generator
- An electromechanical pusher (this is used to drive the pendulum at the start of the wave machine)

If you want to make your own wave machine, use a slightly more than 2-meter length of duct tape, onto which you place barbeque skewer sticks 5 cm apart. Put a layer of duct tape over the top to make the sticks secure. Skewer fruit pastilles (or any other soft sweet) onto the ends of the skewers. Then suspend with one end firmly fixed and the other tied to a string that goes over a pulley to a tensioning weight. If the sticks are not all horizontal when at rest, adjust the sweets along the sticks to obtain the right balance.

**FIGURE 26.1**
The experiment apparatus.

## The Variables

The independent variable is the tension in the string and the tape, and the dependent variable is the speed of the wave.

## The Physics

The important relationship is that between frequency, wavelength, and speed—which, as previously discussed, is given by:

$$v = f\lambda$$

We will use this to calculate the speed of the wave, by observing the wavelength of the wave and the frequency as set on the generator. We do this by observing 'standing' waves. When a wave reaches the fixed end, it is then reflected back along its original path. It then combines with the original wave, either strengthening or weakening it at various positions. At particular frequencies it will produce a standing wave—one that looks perfectly stationary. Figure 26.2 shows what a standing wave of five half-wavelengths looks like.

The tape, string, or wire is not actually stationary; it is oscillating between the two continuous lines shown in Figure 26.2, so it looks like it is at rest. The significance of the standing wave for this experiment is that we can count the number of half-wavelengths and hence the number of full wavelengths and then, using the frequency of the generator, we can calculate the speed of the wave.

We can find the theoretical speed of a wave on a string using Newton's second law fairly straightforwardly. Consider a segment of the string of length $\Delta s$ at the top of the propagating wave, as shown in Figure 26.3.

Although this is moving, it is moving at a constant speed, so Newton's second law is still applicable. There is some radius of curvature at this point, say $R$. The acceleration of this segment towards $O$ is given by $v^2/R$.

Now consider the forces acting on the segment: $T$ is the tension in the string and the force downwards on the segment. $F$ is the sum of the vertical components of the tensions; see Figure 26.4.

**FIGURE 26.2**
A standing wave of five half-wavelengths.

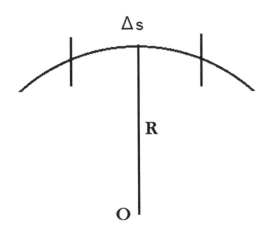

**FIGURE 26.3**
A small part of a string carrying a wave.

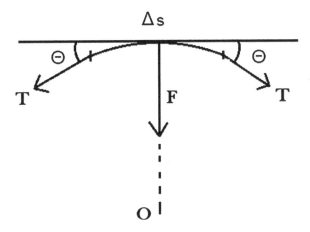

**FIGURE 26.4**
The forces on a small part of a string carrying a wave.

Resolving vertically, the equation of motion is:

$$2T\sin(\theta) = mv^2/R$$

As $\theta$ is small, $\sin(\theta)$ can be approximated by $\theta$. The mass of the segment is given by its mass per unit length ($\mu$) times its length:

$$m = \Delta s\, \mu$$

Where $\Delta s = 2R\theta$. Substituting into the previous equation gives us the mass as:

$$m = 2R\theta\,\mu$$

Substituting this into the equation of motion yields:

$$2T\theta = 2R\theta\,\mu\,v^2/R$$

Which can be simplified by canceling identical terms to:

$$T = \mu\,v^2$$

Which can be rearranged to give the speed as:

$$V = (T/\mu)^{1/2}$$

## The Method

Switch the frequency generator on and slowly increase the frequency until you see a standing wave like that shown in Figure 26.5.

This shows five half-wavelengths. You should note the frequency at this point. Repeat using different weights, recording the number of half-wavelengths and the frequency for each. You need to take great care to spot the best point for the standing waves for each reading.

## The Simulation

The frequency generator can be switched on by clicking the on/off button. Use the Frequency knob to adjust the frequency either by using the mouse wheel or by dragging when using the on-screen controls. The wave machine is 2 m in length. The weights start at 0.5 kg. You can increase and decrease the weights by using the mouse wheel or by clicking when using on-screen controls. Each of the additional weights is 0.25 kg.

**FIGURE 26.5**
A standing wave.

## The Results

Create a table showing your results like this:

|  | 0.5kg | 0.75kg | 1kg | 1.25kg | 1.5kg |
|---|---|---|---|---|---|
| Half-wavelengths | 4 |  |  |  |  |
| F/Hz | 1.18 |  |  |  |  |
| Wavelength/m | 1 |  |  |  |  |
|  |  |  |  |  |  |
| Speed | 1.18 |  |  |  |  |
|  |  |  |  |  |  |
| Speed squared | 1.3924 |  |  |  |  |

Plot the speed squared on the Y-axis and the weight on the X-axis and you should see a linear relationship, showing that the square of the speed of propagation is proportional to the tension.

## Further Discussion

Can you deduce from the Physics what the speeds you inferred from your observations would be if each pendulum was twice the weight?

# 27

## Measuring the Speed of Water Ripples

### Introduction

A water ripple is a wave that propagates in water. It is a vertical oscillation of the water. We've all seen rain falling onto the surface of a still pond or large puddle. Each raindrop that falls creates a circular wave that moves outwards from the point of contact of the raindrop and the pond. The wave crest travels outwards at constant speed. These waves are a little unusual as there is only one created for each wave, whereas most waves continuously emit crests; for example, the waves breaking onto a shore. In our experiment, we need a continuous supply of crests, and we can arrange this by using a plunger that disturbs the water at regular periods. Like the waves on a rope, this is an example of a transverse wave where the direction of the wave is at right angles to the disturbance, which is up and down. If necessary, familiarize yourself with the basic terminology and relationships of a wave by reading the first section of Chapter 26.

### The Objective

To determine the speed of waves in water.

### The Apparatus

You will need:

- A ripple tank
- An electrical plunger
- A signal generator
- A strobe light with frequency controller
- A white screen slightly wider than the length and width of the ripple tank
- A ruler

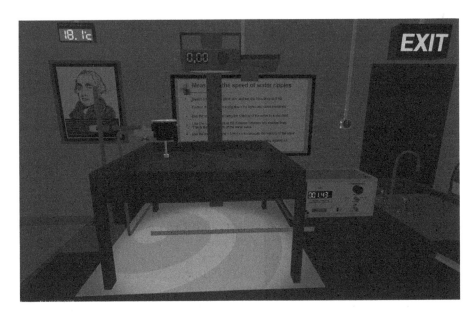

**FIGURE 27.1**
The experiment apparatus.

The signal generator should be connected to the plunger that just touches the water at full extension. The strobe should be placed above the ripple tank so that the shadow of the water waves is projected onto the screen.

## The Variables

The independent variable is the frequency of the waves, and the dependent variable is the speed of the waves.

## The Physics

The important relationship is that between frequency, wavelength, and speed—which, as discussed in the previous chapter, is given by:

$$v = f\lambda$$

We will use this to calculate the speed of the wave, by observing the wavelength of the wave and the frequency as set on the generator.

## The Method

The basic idea is to set the plunger hitting the water with a given frequency, and then to match this on the strobe so that the projected wave appears to be stationary. This makes the wave's length much easier to measure.

Switch the signal generator on and dim the room lights, which will make the waves' shadows easier to see. Set the frequency to 2.00 Hz and set the strobe to the same. Measure the distance between two crests and record your results in the first row of this table:

| Frequency/Hz | Pos1/cm | Pos2/cm | Wavelength/cm | 1/$\lambda$ cm |
|---|---|---|---|---|
| 2 | 36 | 63 | 27 | 0.037037 |
| 3 | | | | |
| 4 | | | | |
| 5 | | | | |

Repeat the same process for all the frequencies given in the table and record your results.

## The Simulation

Lower the lights in the room using the light switch. You can draw the blinds using the switch on the wall beneath the window. Switch on the frequency generator using the red button. Use the scroll wheel on the knob marked 'Frequency' to change the frequency of the plunger. Match the frequency on the strobe using the red knob on the strobe.

Position yourself in a good place to see the wave shadows and the ruler. You will almost certainly need to zoom in (use plus and minus on the keyboard or the Z and X keys. If you are using on-screen controls, then there are magnifying icons on the screen that you can click on). You may need to adjust the light level so that you can see the wave shadows and the ruler markings at the same time.

## The Results

You can use the formula:

$$v = f\lambda$$

To calculate the velocity of the wave at each frequency, these should all be approximately the same. You should plot the frequency on the Y-axis and one over wavelength on the X-axis for all the results. You should get a straight line where the gradient is the required velocity.

## Further Discussion

There is a problem with the experiment in that the ruler is calibrated correctly in centimeters, but the wave shadow is in fact an enlargement due to the projection by the strobe light. You should be able to think of several different ways of adjusting for this. Try to think of a physical way of adding to the apparatus so that the distance can be measured accurately. How would you compensate for the enlargement by using geometry alone?

Note: There is a ruler at the back of the apparatus that should help you get the values you need for the geometric adjustment to be made.

# 28

## Infrared Radiation

## Introduction

Infrared radiation is the radiation from the part of the electromagnetic spectrum just below visible light.

We feel the energy from this radiation as heat. We can feel this quite strongly when we face the sun but perhaps not so much in winter. We can also feel it from objects that are nowhere near as hot, such as a central heating radiator—which, due to the heating fluid being water, can never be more than 100 °C. A black surface is a better emitter and absorber than a white surface, which is one reason why tarmac can get very sticky in the summer sun and we readily feel the heat coming off a hot tarmac road. The tarmac is absorbing large amounts of the solar radiation and emitting it back to us.

## The Objective

To determine which surface from a selection of surfaces emits the most energy.

## The Apparatus

You will need:

- A Leslie cube; this is a cube that can be filled with hot water and has four different surfaces

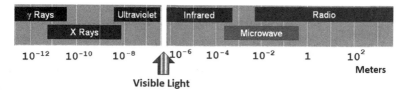

**FIGURE 28.1**
The electromagnetic spectrum.

DOI: 10.1201/9781003262350-29

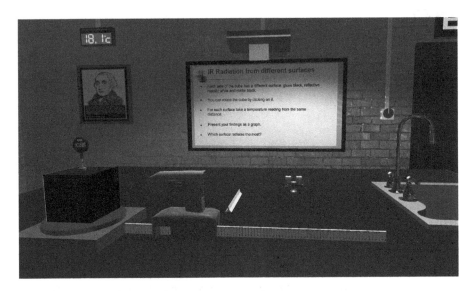

**FIGURE 28.2**
The experiment apparatus.

- An infrared detector; an infrared blood temperature thermometer that does not need to be held against a person can be used
- A ruler

The apparatus should be set up so that the thermometer points to the center of one side of the Leslie cube. The Leslie cube can optionally have a thermometer inserted so that the water temperature can be monitored.

## The Variables

The independent variable is the type of surface that we are measuring the radiation from. The dependent variable is the temperature reading from the thermometer, which is proportional to the energy being emitted from the surface of the cube.

## The Physics

All objects emit and absorb infrared radiation. The hotter the object, the more it will radiate. When an object is hotter than its surroundings, it will emit more radiation than it absorbs and therefore it will cool down; that is, its temperature will drop. Objects will either heat up or cool down until they are in equilibrium with their surroundings; at this point, their temperature becomes stable and they will be emitting the same amount of energy as they are absorbing. Energy is emitted and absorbed from the surface of a body, and the nature of the surface has a big impact on how efficient the body is at being an absorber and an emitter.

## The Method

Fill the Leslie cube with hot water and wait a few minutes until all the surfaces have reached the same temperature. This can be checked by holding a thermometer against each surface. Place the thermometer at a fixed distance—say, 10 centimeters from the surface of the Leslie cube—and read the temperature. Note the surface facing towards the thermometer and the temperature. Repeat this for the other three surfaces as quickly as possible to prevent the water in the Leslie cube from cooling by more than a very small amount.

## The Simulation

Move the thermometer on the ruler to the 10-cm mark; there is a pointer on the bottom of the wooden base that is positioned at the same horizontal position as the sensor to aid this. Position yourself so that you can see the thermometer readout; you may need to zoom in to get a good enough view. You need to click on and hold the red button for several seconds to take a reading. Click on the Leslie cube to rotate it.

## The Results

You should find that the matte black surface is the best emitter. The worst surface at emitting is the metallic surface.

## Further Discussion

Can you think of any way to compensate for the temperature of the water dropping slightly between the readings for each surface?

Find out what is meant by black body radiation.

Why do you think there is no such thing as a perfect black body in nature?

How do we achieve a good approximation to a perfect black body?

# 29

## Diffraction Using a Monochromatic Laser

### Introduction

Light behaves as a wave in this scenario and as such, it is capable of interference in much the same way as water waves can interfere. Peaks and troughs cancel each other, two peaks make a bigger peak, and two troughs make a deeper trough. In particular, if light is passed through a grating, there will be a pattern of constructive and destructive interference. It is that pattern from a grating, where we know the number of lines per millimeter, that is going to be measured in this experiment to determine the wavelength of the light emitted by the laser. Once this has been found, we can then determine the number of lines per millimeter for a different grating as an additional exercise.

### The Objective

To determine the wavelength of the laser and, once this has been found, to determine the number of lines per millimeter for a different grating.

### The Apparatus

You will need:

- A laser source – a cheap laser pointer is ideal
- Two 1.2-meter or longer rulers
- Two diffraction slides with different line spacings
- Wooden mounts for the ruler, slides, and laser

Configure the apparatus as indicated in Figure 29.1. Lasers should be handled with great care. Never point a laser light at another person or randomly. Always know where the light is going before switching it on. Users should be aware of official safety advice such as that available from Consortium of Local Education Authorities for the Provision of Science Services (CLEAPSS).

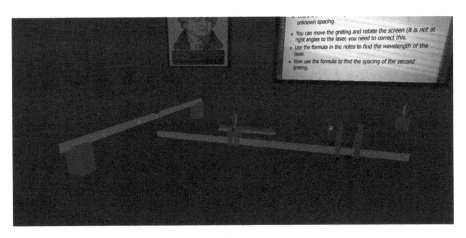

**FIGURE 29.1**
The experiment apparatus.

The laser source is on the right, positioned on the ruler. The diffraction grating is also on the ruler near the middle of the screen. The ruler on the left is positioned to intercept the diffracted light.

## The Variables

The independent variable is the distance from the grating to the screen (which is also the ruler). The dependent variable is the distance from the zero order to the second order maximum.

## The Physics

Consider Figure 29.2, which shows light from two different slits in the grating arriving at the same spot on the screen, where the grating and the screen are at a distance of $L$ and the slits are a distance $d$ apart.

In the diagram in Figure 29.2, we see the paths of two light rays starting a distance of $d$ apart. As $d$ is very much smaller than $L$, we can consider that the beams are for all intents and purposes parallel, which means that the difference in length is $d\sin(\Theta)$ (from the small right-angled triangle with hypotenuse $d$). For the wave from each of these rays to constructively interfere, they need to be an exact multiple of a wavelength apart. So we can write:

$$d\sin(\Theta_m) = m\lambda \text{ (equation 1)}$$

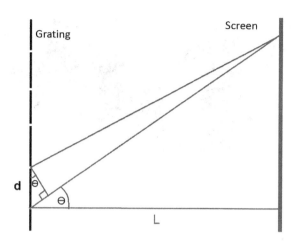

**FIGURE 29.2**
The geometry of the grating and the screen.

Where *m* is the number of the order. In the diagram, the bright spot in the middle is the zero order maximum, the two to either side are the first order maxima, the next two are the second order maxima, and so on.

We can denote the distance from the zero order maximum to the *m*th order maximum as $y_m$. From the diagram, using the big triangle with base L:

$$\sin(\Theta_m) = y_m/(L^2 + y_m^2)^{0.5}$$

Remember that the sine of an angle is the opposite over the hypotenuse of a right-angled triangle. We have calculated the hypotenuse using Pythagoras's theorem.

If we now substitute this into equation 1, we get:

$$dy_m/(L^2 + y_m^2)^{0.5} = m\lambda$$

or:

$$\lambda = (d/m)y_m/(L^2 + y_m^2)^{0.5} \text{ (equation 2)}$$

You should use the second order (m = 2) and obtain the readings for L and $y_2$. You can calculate *d* from what is written on the grating.

- **Question**: What is the highest order maximum visible in the screenshot?

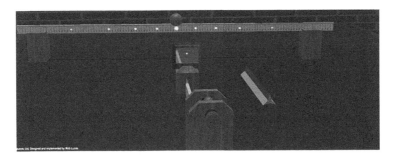

**FIGURE 29.3**
The orders of the diffraction pattern.

## The Method

Switch the laser on and switch off the room lights, which will make the diffraction order maxima much easier to see.

Rotate the screen ruler so that the second order maxima are at an equal distance from the zero order maximum. This makes sure that the screen ruler is perpendicular to the laser light source.

Adjust the position of the grating so that the orders are well spread out along the ruler, which is acting as a screen.

Take measurements of the distance between the grating and the screen, $L$, and then between the zero order and the second order maximum using the screen ruler.

## The Simulation

The ruler on the left is initially not quite at right angles to the zero order light beam, and you will need to rotate it into perfect alignment by using the mouse wheel when the cursor is over the central red knob.

You can switch the laser on using the red knob on the top.

You can switch off the room lights using the switch by the door.

The grating originally on the ruler is marked with the number of lines per millimeter. You will need this for your calculation.

The central grating can be moved along the ruler by placing the mouse cursor over it and using the mouse wheel. The exact position of the grating can be determined from the pointer, as shown in Figure 29.4.

You can zoom in and out using the plus and minus or Z and X keys on the numeric pad.

Clicking on the other grating, which is to the back and right of the table, will swap it with the 500 lines per millimeter grating.

The laser needs to be off for the swap to occur.

**FIGURE 29.4**
Positioning the grating.

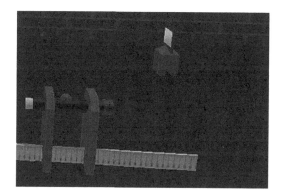

**FIGURE 29.5**
The second grating.

## The Results

Using equation 2:

$$\lambda = (d/m)y_m/(L^2 + y_m^2)^{0.5}$$

Substitute your values for $L$ and $y_2$. You can determine $d$ from what is written on the grating.

## Further Discussion

Now that you know the wavelength, you can calculate the slit spacing of any other grating.

With a different grating of unknown spacing in position, measure the distance to the furthest order maximum you can see. We can rearrange equation 2 to allow us to calculate $d$:

$$d = m\lambda(L^2 + y_m^2)^{0.5} / y_m$$

When you have the spacing, you need to use it to calculate how many lines per millimeter, which is given by:

$$\text{Lines per mm} = 0.001/d$$

# 30

## Inverse Square Law for Gamma Radiation

### Introduction

All forms of radiation follow the inverse square law; that is, the intensity of radiation declines as to the square of the distance from the source. This can be mentally justified quite simply by considering the relative areas of concentric spheres. The area of a sphere is given by:

$$A = 4\pi R^2$$

This tells us how the area of the 'front' expands from the origin. If it needs to cover an area proportional to $R^2$, then it is going to be weakened or attenuated by $1/R^2$, as the same amount of radiation now has to cover an area that has increased by a factor of $R^2$.

- **Question**: What is the area that a light source illuminates after 1 millionth of a second, 2 millionths of a second, and 3 millionths through to 10 millionths. Draw a graph of time on the X-axis against area on the Y-axis.

The astronomer Johannes Kepler appears to be the first to argue (in 1604) that the intensity of light from a source obeys the inverse square law.

Gamma rays are electromagnetic radiation – the same as light and radio waves, but at a high frequency, which means a high energy. In fact, they are the highest-frequency form of electromagnetic radiation observed. They were first discovered by French chemist Paul Villard while he was investigating the radiation from the element radium. It was our old stamp-collecting friend, Ernest Rutherford, who named them gamma rays.

In this experiment, we measure the background radiation and the count rates of gamma particles hitting a detector at a range of distances. Plotting the results will verify the inverse square law.

### The Objective

To show that gamma radiation follows the inverse square law, in that the intensity falls off in proportion to one over the square of the distance from the source.

DOI: 10.1201/9781003262350-31

## The Apparatus

You will need:

- Safe source of gamma radiation such as cobalt-60
- Geiger counter
- 1 meter rule
- Thick lead block
- Stopwatch

The apparatus should be assembled as shown in Figure 30.1. Teachers, technicians, and students must be aware of all regulations involved in handling radioactive substances. Cobalt-60 is a common radioactive substance often used in this experiment, and it must be handled in accordance with these regulations. Naturally, there are no safety concerns with using the simulation.

The counter on the left of Figure 30.1 counts the number of gamma rays hitting the detector that is to the right of it. The counter and detector are connected by the cable. To the right of the detector is a ruler. At the far right is the gamma ray source. There is also a large lead block that can be used to block the gamma rays.

## The Variables

The independent variable is the distance, and the dependent variable is the incident count rate.

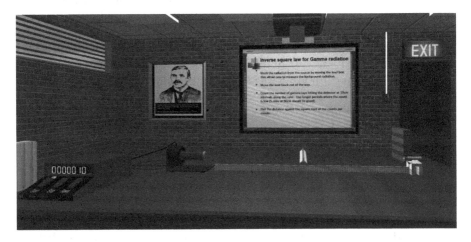

**FIGURE 30.1**
The experiment apparatus.

## The Physics

An account of the reasoning behind the inverse square law is given in the introduction to this chapter.

## The Method

Start by moving the lead block between the ruler and the source as close to the source as possible. Start your timer at the same time as you press the start button on the electronic counter. Stop the counter at 20 minutes.

Now remove the lead block and move the source to 60 cm on the ruler. Count the particles for a 10-minute period. Repeat this for distances at 10-cm intervals down to 10 cm. You can time for less than 5 minutes where the counts are high. Do not forget to record the actual period length.

## The Simulation

The gamma ray source can be moved by placing the mouse cursor on it and using the scroll wheel.

You can move the lead block between the ruler and the source by putting the mouse cursor over the block and using the mouse scroll wheel. This will cut off the gamma rays moving in the direction of the detector and will enable you to get a good reading for the background radiation count. You may need to move the source to the right to enable the movement of the lead shield.

You can drop down to the level of the table by using the page down key, and you can zoom in and out on the apparatus using the plus and minus keys. You will find this useful when accurately placing the source on the ruler.

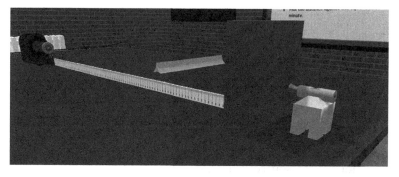

**FIGURE 30.2**
The radiation source behind the lead block.

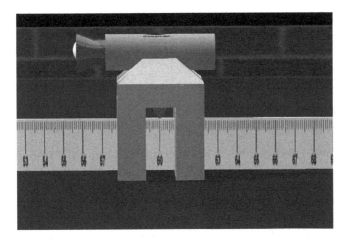

**FIGURE 30.3**
Reading the position of the source.

Notice the pointer in the source carrier in Figure 30.3 that enables you to accurately position the source when used with the zoom feature.

## The Results

Create a table that has columns for the time period, the total count, the count rate, the corrected count rate (taking account of the background radiation), the distance, and 1 over the root of the count rate.

| Time /mins | Total Count | Count/min | Corrected Count/min | Distance cm | 1/ √(Corrected Count Rate) |
|---|---|---|---|---|---|
| | | | | | |
| 5 | 490 | 98 | 86.1 | 60 | 0.10777 |
| 5 | | | | | |
| 4 | | | | | |

When you have found all the values, plot the graph of the last two columns, which should be a straight line.

## Further Discussion

Why does your straight line not go through the origin? Do we know the exact positions of the source and the detector?

# 31

## Refraction of Light

### Introduction

When light passes from one medium to another, it is refracted – or bent – at the boundary between the two mediums. This can be easily observed by looking at a drinking straw in a glass of water. The straw looks to be bent at the contact surface between air and water. The amount by which the light is bent is different according to a property of the medium called the refractive index. The higher the refractive index, the more that light is bent. Diamond has a very high refractive index, and it is this property that allows many facets to be ground onto a diamond, making it sparkle much more than can be achieved with glass.

The exact law of refraction is called Snell's law of refraction, after the Dutch astronomer Willebrord Snellius (1580–1626). The law gives the relationship as:

$$\sin(i)/\sin(r) = n1/n2$$

Where $i$ is the incident angle, $r$ is the refracted angle, $n1$ is the refractive index of the first medium, and $n2$ is the refractive index of the second medium.

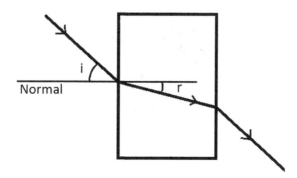

**FIGURE 31.1**
A glass block with incident and refracted rays.

DOI: 10.1201/9781003262350-32

## The Objective

To demonstrate that different mediums bend light by different amounts and to confirm Snell's law of refraction.

## The Apparatus

You will need:

- A light box
- A collection of rectangular blocks of different transparent mediums
- Some graph paper
- A protractor

Arrange the glass block to be in the path of the light from the light box, both placed on top of a large sheet of graph paper so that the light ray is traveling along one of the axes of the graph paper.

## The Variables

The independent variable is the incident angle. The dependent variable is the refracted angle.

## The Physics

Light is a wave that travels at $c$, the speed of light, in a vacuum. In any other medium, light travels at a slower speed. The refractive index of a medium is directly connected to the speed of light in that medium. Think of wavefronts

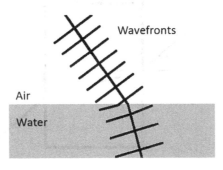

**FIGURE 31.2**
Wavefronts meeting a medium of slower transmission speed.

of light arriving at the surface of a substance like water. As a wavefront reaches the water, it travels more slowly and so bends the light towards the normal.

If this is difficult to imagine for a light wave, imagine that each wavefront is a line of soldiers marching towards the water's edge at an oblique angle. As the first soldier reaches the water, he slows down, and the rest of the soldiers carry on at the same speed until the next soldier reaches the water and slows down. The overall effect is to pivot the soldiers' direction towards the normal. Snell's law can be restated as:

$$\sin(i)/\sin(r) = v1/v2$$

Where *i* is the incident angle; *r* is the refracted angle, as before; *v1* is the speed of light in the first medium; and *v2* is the speed of light in the second medium. This can be derived in a variety of ways. However, the geometric construction based on the earlier observation is very simple. Consider a light front arriving so that the left part just reaches the second medium at point A, as shown in Figure 31.3.

At some time, *dt* later, the right-hand part of the wavefront reaches the medium. Then:

$$AC = v2dt$$
$$BD = V1Dt$$

Simple geometry tells us that the angle ADC = r, and the angle BAD = i. Therefore, considering the right-angled triangles ADC and BAD:

$$\sin(i) = V1dt$$
$$\sin(r) = V2dt$$

From which the law can be directly derived.

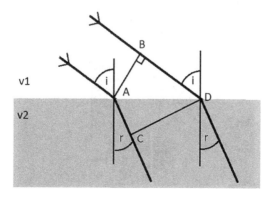

**FIGURE 31.3**
Wavefronts meeting a medium of slower transmission speed.

## The Method

For demonstrating that different substances bend light by different amounts, simply use the protractor to measure these angles.

To verify Snell's law, proceed as follows. Set the incident angle for the light ray to 20 degrees. You can do this by having the glass block on graph paper and aligning the paper with the beam of light. Then, by putting the protractor up to the edge of the block as shown in Figure 31.4, you can measure the incident angle by looking at the line (vertical in the picture) from the center of the protractor to its edge. Figure 31.4 shows this angle being set to 30 degrees; this is the incident angle, $i$.

You can either use a protractor directly to measure the refracted angle or use a pencil to mark the entry and exit points of the ray on the graph paper, then remove the block and create the line joining the points you have just made, and then measure the angle with the protractor. This is the refracted angle, $r$. Repeat this for a range of angles, noting the incident angle and the refracted angle for each case.

## The Simulation

This simulation is a little different from all the others, as you have an overhead view and cannot move around all of the laboratory. You can move a restricted amount by using the cursor keys or the onscreen left joystick control. You will find it useful to use the magnifier to get up close to the block and protractor to take accurate measurements of the incident and refracted angles.

Each block can be selected by clicking on it; the block will then be moved to the graph paper immediately in front of the light box. The blocks, once in

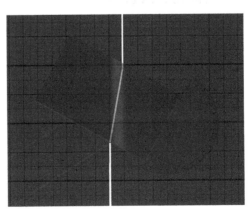

**FIGURE 31.4**
Measuring the angle of incidence with a protractor.

place, can be rotated either by using the mouse wheel or by dragging when using the onscreen controls.

You can position the protractor by dragging it; it can be rotated using the mouse wheel or the right-hand joystick when using the onscreen controls.

## The Results

Create a table of your results and calculate the sin(i)/sin(r) value. Compare this with the known values for the refractive indices of the various transparent materials.

| I (deg) | r (deg) | sin(i) | sin(r) | sin(i)/sin(r) |
|---------|---------|--------|--------|---------------|
|         |         |        |        |               |
| 30      | 20      | 0.5    | 0.34202 | 1.4619022    |
|         |         |        |        |               |
|         |         |        |        |               |
|         |         |        |        |               |

## Further Discussion

Not all the light from the beam is refracted; there will usually be some light that is reflected, as you can see in Figure 31.5. Confirm that the angle of incidence is always equal to the angle of reflection by measuring these for different angles using different blocks.

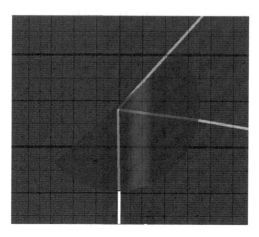

**FIGURE 31.5**
Internal reflection.

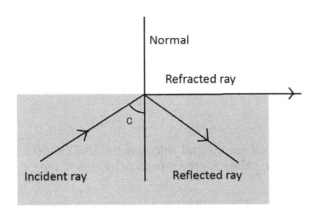

**FIGURE 31.6**
Schematic showing total internal reflection.

Total internal reflection occurs when the light ray inside the material, such as glass, is completely reflected internally.

In Figure 31.6, you can see the refracted ray just skimming the top of the semicircular glass block, while the reflected ray behaves normally, with its incident and reflected angle the same.

Using the semicircular block, find the angle at which there is total internal reflection. This is called the critical angle, and it is marked by $c$ in Figure 31.6.

From Snell's law applied to Figure 31.3, we get:

$$\sin (90)/\sin(c) = n \text{ (the refractive index)}$$
$$n = 1/\sin(c)$$

Verify that this formula is consistent with the value you measured.

# 32

## Magnetic Field Due to a Coil of Wire

## Introduction

Hans Christian Ørsted discovered in 1819 that current flowing in a wire created a magnetic field. By 1820, Jean-Baptiste Biot and Félix Savart discovered the Biot–Savart law, which allows the calculation of the magnetic field produced by any wire that is carrying a current.

This setup allows for several experiments that measure the magnetic field due to a coil of wire. It is possible to establish the relationship between:

- The number of turns of the coil and the magnetic field strength
- The effect of the current in the coil on the magnetic field strength
- The strength of the field along the radial axis of the coil

It is also possible to measure the horizontal component of the Earth's magnetic field in the laboratory.

## The Objective

To find the relationships between the current in a coil, the number of turns in a coil, the axial position of a coil, and its magnetic field.

## The Apparatus

You will need:

- A magnetometer (a magnetometer is really just a magnetic compass with a different needle arrangement to make measuring deflections easier)
- A coil of wire with various taps to a varying number of turns
- A pedestal/ramp that allows for the radial motion of the magnetometer
- A ruler
- A power supply unit (PSU)
- A multimeter that is set to display a current of up to 10 amps

DOI: 10.1201/9781003262350-33

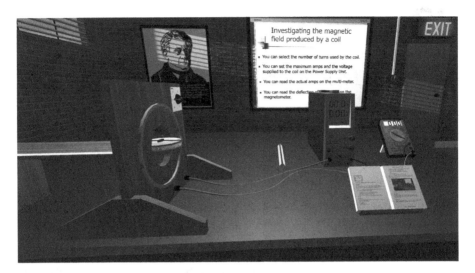

**FIGURE 32.1**
The experiment apparatus.

The apparatus should be oriented so that the needle of the magnetometer is pointing at right angles to the coil when there is no current in the coil. This is when the magnetometer is aligned with the Earth's magnetic field.

The coil, the PSU, and the meter should be connected in series, as shown in Figure 32.1.

## The Variables

The independent variables are the current, the number of turns in the coil, and the axial position. The dependent variable is the deflection of the magnetometer that indicates the strength of the magnetic field.

## The Physics

When there is current flowing, the magnetic force on the magnetometer's needle is the vector addition of the force from the coil's magnetic field and that of the Earth.

If we denote the magnetic field force of the Earth by $B_{earth}$ and that of the coil by $B_{coil}$, then the resultant force is $B_{total,}$ and the relationship between them is given by:

$$\tan(\Theta) = B_{coil} / B_{earth}$$

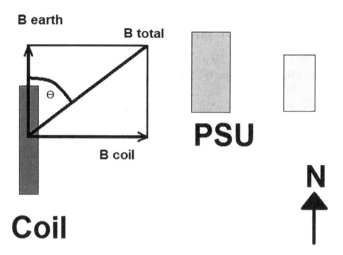

**FIGURE 32.2**
Schematic of the magnetic field directions.

Therefore:

$$B_{coil} = B_{earth}\tan(\Theta)$$

Therefore, to show the relationship between the current and $B_{coil}$, we need to plot the current against the tangent of the deflection.

## The Method – Current and Magnetic Field

For demonstrating the relationship between the current and the magnetic field, we proceed as follows:

- Turn on the output of the PSU and set its maximum current to 5 amps
- Now select two windings for the coil
- Take deflection readings for the current at various values, from about 0.2 amps through to 4 amps, by changing the voltage and using the current reading on the multimeter

## The Method – Number of Turns and Magnetic Field

With the same setup up as in the previous section, take deflection readings for the different number of turns available (1, 2, 3, 5, and 7) using the same current. Record the deflections.

## The Method – Radial Distance and Magnetic Field

Set the number of turns to 10 and the voltage to a sufficient value to get a reading of the magnetometer of over 60 degrees. Measure and make a note of the radius of the coil.

Now move the magnetometer in 2-cm stages, noting the deflection at each stage.

## The Simulation

North–south is aligned with the left and right laboratory walls, with north being in the direction of the projector screen. Note that the north magnetic part of the needle of the magnetometer is in the stubby, central part of the needle and pointing to the left in Figure 32.3; this is what makes reading the deflections easier.

The means of operating switches and knobs is to place the mouse cursor over the object – this will cause it to change color to yellow – and then either left-clicking with the mouse (for a switch) or using the mouse wheel (for adjusting the position of a rotating knob).

The coil has several taps that can enable a variable number of turns to be investigated.

The maximum current can be set using the knob marked 'CURR' on the PSU, and the voltage can be altered using the knob labeled 'VOLT'. The switch marked 'Output' is used to start and stop the current output from the PSU.

**FIGURE 32.3**
The magnetometer.

**FIGURE 32.4**
Setting the number of turns of the coil.

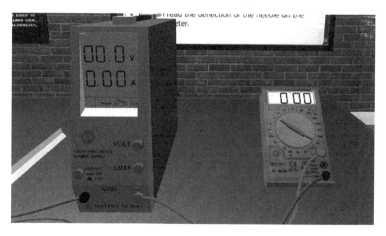

**FIGURE 32.5**
The PSU and multimeter.

The radius of the magnetometer is 3 centimeters, so it may be helpful to put a column into your table for the radial experiment that indicates where the leading edge of the magnetometer is for each *d* value.

The radius of the coil is 0.08 meters.

## The Results

To demonstrate the relationship between the current and the magnetic field, plot the current against the tangent of the recorded deflections. This should give a straight line, showing that the magnetic field of a coil is directly proportional to the current through the coil.

To demonstrate the relationship between the number of turns and the magnetic field, plot the number of turns against the tangent of the deflections. You should obtain a straight line. This indicates that the magnetic field strength of a coil is directly proportional to the number of turns in the coil.

To demonstrate the relationship between radial distance and magnetic field, your first few entries might be:

| d /m | | Theta /deg | | tan(theta) |
|------|---|------------|---|------------|
| 0.02 | | 65 | | 2.143309 |
| 0.04 | | 59 | | 1.663547 |
| 0.06 | | 50 | | 1.191355 |
| 0.08 | | 40 | | 0.838875 |

Plot the radial distance against the tangent of the angle of deflection. Remember that this tangent value is proportional to the magnetic field and not an actual value of the field.

## Further Discussion

The experimental setup also allows us to calculate the horizontal component of the Earth's magnetic field.

We can calculate this by using the formula for the magnetic field caused by a coil:

$$B_{coil} = \mu_0 Ni/2R$$

Where $\mu_0$ is a constant called the permeability of free space and has the value:

$$\mu_0 = 4\pi \ 10^{-7} \ TmA^{-1}$$

$R$ is the radius of the coil. If we divide both sides of this equation by $B_{earth}$, we get:

$$B_{coil} / B_{earth} = \mu_0 Ni/2RB_{earth}$$

$B_{coil} / B_{earth}$ is $tan(\Theta)$. The gradient of our graph from the first experiment is $tan(\Theta)/i$.

Therefore, if we solve the equation:

$$\tan(\Theta)/i = \text{gradient from first graph} = \mu_0 \, N/2RB_{earth}$$

for $B_{earth}$, we will find its value:

- Calculate the gradient from the first experiment
- Rearrange the equation given earlier at $B_{earth} =$
- Substitute the values for gradient, $\mu_0$, $N$, and $R$

For the simulation, you should get a value somewhere in the region of $2.8 \times 10^{-5}$ T.

# 33

## Investigation of Magnetic Flux of a Current-Carrying Wire

## Introduction

This experiment determines the field strength of a magnetic field by observing the force it exerts on a current-carrying wire.

## The Objective

To find how the strength of a magnetic field due to a wire varies with the current through the wire.

## The Apparatus

You will need:

- A power supply
- Copper wire
- Digital kitchen-type scales with zero function
- Multimeter
- Wires and crocodile clips
- Two magnets in a U-shaped yoke
- Wooden mounting blocks

The power source is connected in series with an ammeter and a copper wire.

The copper wire is supported between an arrangement of magnets in a U-shaped yoke, which provides a close approximation to a uniform field. The yoke containing the magnets sits on the scales, which measure the force on the yoke from the magnetic field.

DOI: 10.1201/9781003262350-34

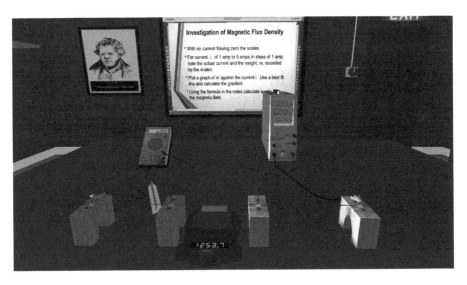

**FIGURE 33.1**
The experiment apparatus.

## The Variables

The independent variable is the current. The dependent variable is the strength of the magnetic field as measured by the scales.

## The Physics

The force on a single charge $q$ moving at speed $v$ due to a magnetic field $B$ at right angles to the movement of the charge is given by:

$$F = qvB$$

If there are $n$ charges per unit volume in our wire of length $L$ and cross-sectional area $A$, then the total number of charges $N$ is given by:

$$N = nLA$$

Therefore, the total force will be given by:

$$F = NqvB = nLAqvB$$

The current in a wire $i$ is given by:

$$i = nqvA$$

This is substituted into the previous equation to give:

$$F = BiL$$

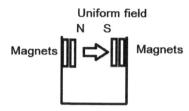

**FIGURE 33.2**
The magnetic field between the magnets.

Where $B$ is the magnetic field, $L$ is the length of the wire in the field (which is the length of the magnets), and $i$ is the current. Therefore, the magnetic field is given by:

$$B = F/Li$$

When you plot the mass registered on the scales against the current, the gradient of your graph gives you a value for $m$ over $i$. Consider forces on the wire. The force from the magnetic field upwards as measured by the scales is $BiL$.

The force downwards on the wire due to gravity is:

$$F = mg$$

The forces must balance, as the wire does not move; therefore:

$$BiL = mg$$

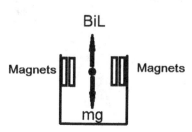

**FIGURE 33.3**
The forces acting on the wire.

Therefore, the result for *B* is given by:

$$B = mg/iL$$

The gradient of the graph, mass against current, is *m/I*; substituting this into the previous equation gives you a value for the magnetic field.

## The Method

The electronic scales indicate the weight of the yoke and magnet assembly. We want to ignore this weight and concentrate on the force due to the current in the wire and the magnetic field. Zero the scales. The scales will now only show any additional weight due to the force from the interaction of the magnetic field and the current in the wire. Normally we would expect the wire to be pushed upwards, but the wire is being held rigidly in the two brackets on either side of the scales, so the force will press on the scales and cause a positive reading.

Switch on the power supply. With the power on, set the maximum current to a little over 5 amps. You should now adjust the voltage until the meter reads as close to 1 amp as you can get, as in Figure 33.4.

Note the reading of the amps and the weight indicated on the scales. Now repeat this for currents of 2, 3, 4, and 5 amps.

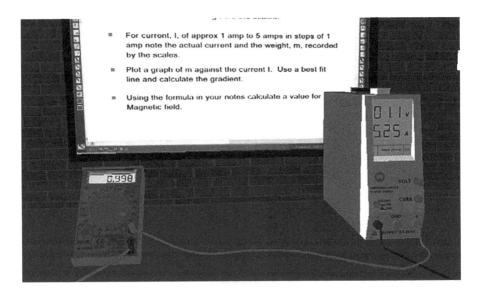

**FIGURE 33.4**
Setting the current to as close to 1 amp as possible.

## The Simulation

You can zero the scales by clicking on the red zero knob.

You can switch on the power supply by clicking the red on/off button on the supply.

The length of the wire is 8.5 centimeters.

## The Results

Complete this table of results:

| m /kg | I /amps |
|---|---|
| 0.0004 | 0.998 |
| | |
| | |
| | |
| | |
| | |

Plot a graph of *m* against *i* and calculate the gradient. Use the formula given in the Physics section to calculate the strength of the magnetic field.

You should obtain a result of around 0.05 Tesla.

## Further Discussion

Would the experiment work as well with a spring balance?

Justify your answer. Hint: think about what might be affected by any movement of the magnets.

# 34

## Magnetic Flux Linkage

### Introduction

Magnetic induction between coils is the mechanism by which transformers work. It was discovered by Michael Faraday in 1831 and is usually known as Faraday's law of induction. James Maxwell correctly quantified the law and made it part of the group of equations known by his name. In Faraday's experiment, he used an iron ring with windings around both sides, as seen in Figure 34.1.

Each time Faraday opened or closed the switch, the galvanometer would register a reading, which did not occur when the switch was kept open or closed. It is the change in the magnetic field that is important and causes the induction. This arrangement is reflected in the modern transformer that we use to change alternating current (AC) voltages by using different numbers of windings on the two coils to either step up or step down the voltage.

This experiment establishes the relationship between the induced voltage in a coil when at various angles to another coil. It uses an audio signal as the varying input voltage.

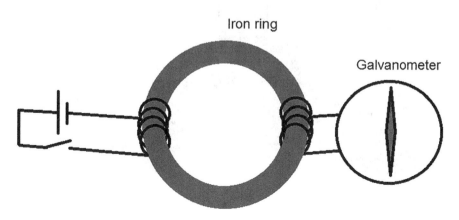

**FIGURE 34.1**
Maxwell's simple apparatus for magnetic induction.

DOI: 10.1201/9781003262350-35

## The Objective

To establish the relationship between the induced voltage in a coil when at various angles to an outer coil.

## The Apparatus

You will need:

- An outer coil of 80 turns
- An inner coil of 40 turns set in the middle of the outer coil and that can be rotated through a measurable angle. This coil is sometimes called the search coil
- An audio signal generator
- An oscilloscope

The outer coil needs to be connected to the output of the signal generator, while the output of the inner coil is connected to the input of the oscilloscope.

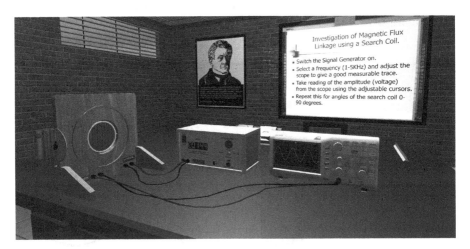

**FIGURE 34.2**
The experiment apparatus.

## The Variables

The independent variable is the angle of the inner coil. The dependent variable is the voltage induced in the inner coil.

## The Physics

The Biot–Savart law gives us the magnetic field due to a small line segment of wire carrying a current $i$ as:

$$dB = (\mu_0/4\pi) \, i \, (dl \times r) \, /r^2$$

In other words, the line segment $dl$ carrying a current $i$ generates a magnetic field $db$ at the point indicated, which is $r$ from the line segment, where $\mu_0$ is the magnetic constant.

The Biot–Savart law can be used to find the equation of the magnetic field along the axis through the middle of the coil (as you are measuring in this experiment). This involves integration, and the workings can be found online. The result is:

$$B(d) = (\mu_0 \, i \, / \, 2) \, N \, R^2 \, / \, (R^2 + d^2)^{(3/2)}$$

You can use this formula to calculate the theoretical values for the magnetic field for the values you have used in your experiment by substituting for $N$ (the number of turns), for $i$ (the current in your circuit), and for $R$ (the radius of the coil).

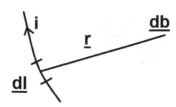

**FIGURE 34.3**
The magnetic field due to a small segment of current-carrying wire.

## The Method

Start with both coils aligned as shown in Figure 34.2. Switch on the audio signal generator and turn up the amplitude. Adjust the vertical sensitivity on the oscilloscope so that the wave fills most of the screen. Adjust the sweep time to fit one or two entire wavelengths onto the screen. You should be looking at something like that shown in Figure 34.4.

You may find that you need to go back and adjust the amplitude to get the best trace. You could measure the voltage by counting the squares, but this oscilloscope allows you to do it in a better way. Different oscilloscopes work in different ways; however, they all basically have the same functionality. The Tektronix oscilloscope shown in Figure 34.4 is common in education and provides a very nice and easy-to-use interface. Obviously, if you are using a different oscilloscope, you are going to need to find out how it is controlled so that this experiment can be followed.

There are two cursors that can be used for measuring voltage. To display these requires you to use some controls that are not particularly intuitive the first time you use them. To start with, you need to press the Cursor button, as indicated in Figure 34.5. This displays the top-level Cursor Menu, as shown in Figure 34.6.

Note the line of buttons alongside the right-hand edge of the screen. These are for selecting menu options that are displayed to the left of the buttons on

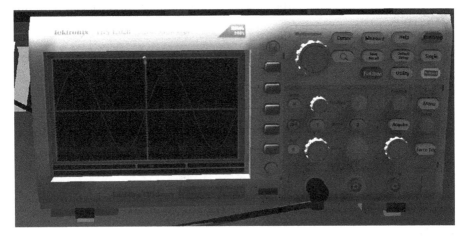

**FIGURE 34.4**
The induced wave as seen on the oscilloscope.

**FIGURE 34.5**
The cursor control on the oscilloscope.

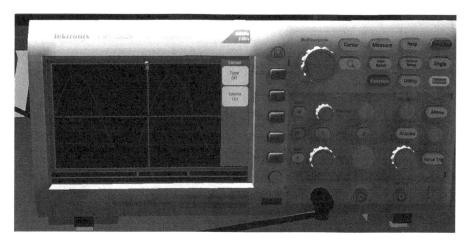

**FIGURE 34.6**
The top-level Cursor Menu.

the actual screen of the oscilloscope. Select the top-most button to display the Cursor submenu as shown in Figure 34.7, indicated by the arrow.

Note that the top-most submenu entry is currently selected, indicating that cursors are switched off.

Rotate the large multipurpose knob just to the right of the menu button you have just used to change the selection to 'Amplitude', as shown in Figure 34.7.

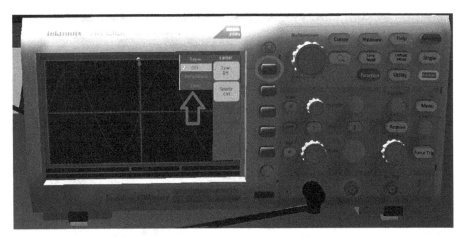

**FIGURE 34.7**
The submenu of the cursor control.

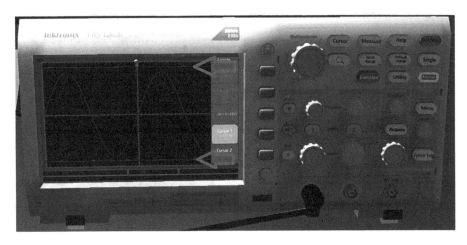

**FIGURE 34.8**
Cursors displayed on the screen.

Now click on the multipurpose button without rotating it. This will cause both cursors to be displayed, as shown in Figure 34.8.

You can adjust the positions of the two horizontal cursors by using the multipurpose knob when the cursor you want to move has been selected from the Cursor menu that is still displayed.

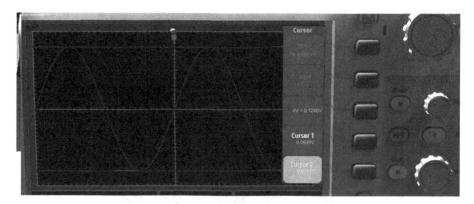

**FIGURE 34.9**
Cursors displayed on the screen touching the wave.

Figure 34.9 shows the cursors when they have been adjusted to align with the maximum and minimum voltages of the trace. Note that the menu entries are indicating the various voltages, the most useful of which is the $dV$ entry. This displays the voltage difference between the top cursor and the bottom cursor. This is the voltage you need to measure in the fashion just described.

Clicking on either of the controls to the right of the menu entries 'Cursor 1' or 'Cursor 2' will select that cursor, which you can then adjust the position of with the large multipurpose knob.

## The Simulation

The signal generator is switched on by clicking on the On/Off button. The frequency can be changed by using the mouse wheel with the cursor over the large red knob. The amplitude of the signal can be changed by using the mouse wheel.

The angle of the inner coil is varied by using the knob at the left central edge of the protractor.

The oscilloscope can be controlled using the mouse buttons and the mouse wheel on the various controls, as described in the previous section.

You will almost certainly need to zoom in to see the voltages clearly (zoom using the +/- or Z/X keys).

Note that not all of the functionality of the Tektronix oscilloscope has been implemented, but it is sufficient for this experiment using the controls described previously.

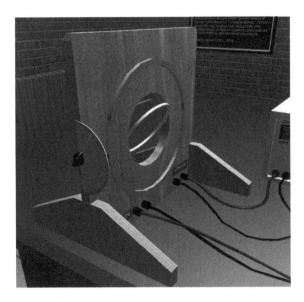

**FIGURE 34.10**
Adjusting the angle of the search coil.

## The Results

Take readings at 10-degree intervals and complete this table:

| Theta deg | Theta Rad | cos(Theta) | V |
|-----------|-----------|------------|---|
| 0 | 0 | 1 | |
| 10 | | | |
| 20 | | | |
| 30 | | | |
| 40 | | | |
| 50 | | | |
| 60 | | | |
| 70 | | | |
| 80 | | | |
| 90 | | | |

When you have completed this table, plot a graph of the cosine of the angle against the voltage. You should get a straight line that indicates that the induced voltage is proportional to the cosine of the angle between the coils.

## Further Discussion

Find out what a Hall probe is and how it works. Then investigate how the same experiment can be done using a Hall probe.

What are the advantages?

Are there any disadvantages?

# Appendix 1

## Uncertainties

### Introduction

All measurements involve some degree of uncertainty as to the accuracy that is achieved in making the measurement. The process of making a measurement is basically observation, where the observer makes the best judgment of what the value is. A value that has been obtained by an observation is generally given in the form:

Value = observed value ± some uncertainty

This recognizes that there is inherently the possibility of some uncertainty in the value. For example:

Length = 0.213m ± 0.001m

Means that the measured length is between 0.212 m and 0.214 m.

We normally round uncertainties to one significant digit, although we might use two digits if the first digit is a 1.

Other factors that contribute to uncertainty include:

- Limitations of the device making the measurement
- Some simplification that has been made in our assumptions, such as assuming there is no friction at all when using the AirTrack
- Environmental changes, such as the atmospheric pressure changing while we are conducting the experiment

## Measurements and Readings

If we are taking a reading from a scale like a ruler, this single measurement is going to have an uncertainty that we estimate to be ± half of one division on the scale. Commonly, a ruler will have millimeters (mm) as the smallest division. So in this case, we take the uncertainty to be ± 0.5 mm. In general, we take the uncertainty to be ± half of the smallest division. This can be thought of as the range (spread) of the measurements, where the same measurement is repeated many times. Given that the scale hasn't been entirely misread, the spread is going to be directly due to the observation of the actual values against the divisions of the measuring device. This is entirely logical. Therefore the range, $R$, is given by:

$$R = x_{max} - x_{min}$$

The associated uncertainty in a measurement is given by half this value:

$$\Delta x = R/2 = (x_{max} - x_{min})/2$$

For a digital readout, such as amps that are displayed to 1 hundredth, the uncertainty could be taken as being ± 0.005A. However, if the accuracy of the instrument is given by the manufacturer, then this should be used.

The astute reader may have noticed that we are taking it for granted that the zero point is completely accurate. This is important, because it is indicating that there are two distinct types of measurement: those that are a difference between two values and those that are a simple reading of one value; this is usually when we can take the zero point as completely accurate. We distinguish these by calling those that derive from a difference as *measurements,* and those that require a single value as *readings*. Therefore, a measurement is the result of two readings, which involves twice the uncertainty of a single reading. Our use of an ammeter needs a single reading, but recording the length of something with a ruler is a measurement requiring two readings.

## Reducing Uncertainty

We obviously want to reduce our uncertainties by as much as we can. The most often used method is to repeat our readings and take the average.

$$x_{avg} = (x_1 + x_2 + \ldots + x_N)/N$$

For example, we might time the descent of a ball bearing three times and take the average. In this case, we divide the uncertainty by the square root of three. The average becomes more accurate as the number of measurements increases. For N measurements, the uncertainty in the mean is:

$$\Delta x_{avg} = \Delta x / \sqrt{N}$$

Sometimes we can take a single measurement of many instances. For example, we may want to know the rate of water flow in liters/second. We can get a better result by taking the measurement of 1 minute and dividing by sixty. Another example is measuring the width of something that is quite small, such as gold foil. We can put ten or more together and measure the width of that. In this case, the uncertainty is divided by N, the number of items being measured.

## Uncertainties in Constants

Uncertainties can also come from the constant values that are given on data sheets, such as $g = 9.81 ms^{-2}$. Here, the uncertainty is taken to be ±1 of the last significant digit. In this case this is ± 0.01 $m/s^{-2}$.

## Combining Uncertainties

When you are adding two values, they must have the same units. For example, we can only add voltages to voltages where both are expressed in volts. Make sure that you are using the same units, then simply add the uncertainties for each value to get the uncertainty of their sum.

When you are multiplying two values – for example, force X distance, to get work done – the units of force and distance are clearly not the same, and so they cannot be simply combined by addition, multiplication, or division. One way of doing this is to convert each to a percentage (which has no units; it is dimensionless). To do this, take each of the readings and express each uncertainty as a percentage. For example:

D = 0.412m ± 0.001m => % uncertainty = (0.002/0.412) X 100 = 0.49%
F = 9.81N ± 0.01N => % uncertainty = (0.02/9.81) X 100 = 0.20%

The combined uncertainty is the sum of the percentages. In this case, the uncertainty in work done is 0.69%.

The work done is = 0.412 X 9.81 = 4.042J ± 41.2 * 0.69/100
= 4.042 ± 0.007 J

For dividing values, we also need to calculate the percentage uncertainties and add them.

A more direct method that is especially useful when plotting the results is to note the value for each variable alongside its uncertainty, then compute the maximum and minimum values using the uncertainty. Then when you multiply the two values, also multiply the maximum and minimum values. This directly gives you the central value and the upper and lower error bars for your graph and sidesteps the need to compute percentages. For example, these three values are in the final three columns of this table:

| F N | ΔF N | Max F N | Min F N | D m | ΔD m | Max D m | Min D m | W J | Max W J | Min W J |
|------|-------|----------|----------|-------|-------|----------|----------|------|----------|----------|
| 9.81 | 0.01 | 9.82 | 9.8 | 0.412 | 0.001 | 0.413 | 0.41 | 4.04 | 4.06 | 4.03 |

This method seems more logical and consistent with our usual requirement to graph the results and can be produced effortlessly when we use a spreadsheet to record the results.

## Math Functions of Uncertainties

Sometimes when processing our results, we apply a mathematical function such as raising to a power, or a trig function such as sine. In these cases, we apply the function to the reading ± the uncertainty. An example is given in the next section.

## In Practice

When we collect our data, we introduce extra columns for the uncertainties and process them according to the rules given earlier. When we plot our results, we use the mean values that we have collected and apply error bars based on our calculated uncertainties. The error bar will extend above and below the mean data point to half the range, which is the value Δ as defined previously.

Where we have a function of a value, such as the use of the trig function tan in the experiment that determines the local magnetic field due to the Earth, we simply take the tan of the observed value plus the uncertainty and the tan of the value minus the uncertainty.

| i amps | Δi amps | Θ deg | ΔΘ deg | $\tan(\Theta)$ | $\tan(\Theta+\Delta\Theta)$ | $\tan(\Theta-\Delta\Theta)$ |
|---------|----------|--------|---------|---------|-------------|-------------|
| 0.26 | 0.01 | 8.00 | 0.5 | 0.141 | 0.149 | 0.132 |
| 0.6 | 0.01 | 18.50 | 0.5 | 0.335 | 0.344 | 0.325 |

This table shows the first few rows of the dataset. The first column is the current, followed by the uncertainty in the current reading in amps. Then we have the deflection measured in degrees, followed by the uncertainty in this measurement. Then we have the tan of this value, followed by the tan of the value plus the uncertainty and the tan of the value minus the uncertainty in theta. This gives us an uncertainty of ±0.009 for the first row. The second row shows a slightly larger value for the uncertainty. This trend will continue for higher values of the current. So when we come to plot the current against the value of tan(Θ), we can indicate the larger uncertainties by using error bars that correspond to those calculated in the table.

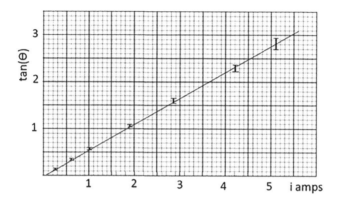

**FIGURE A1.1**
The graph of the results showing error bars.

# Appendix 2

## Using Excel for the Results

## Introduction

Spreadsheets have been one of the most successful programs for personal computers. They are easy to use and extraordinarily powerful. Like most good ideas, the spreadsheet is based on a simple principle. Fundamentally, a spreadsheet is a labeled grid of cells into which we can enter text, numbers, and formulae. The cells are labeled by column and row. The formulae can reference other cells, and it is this that gives a spreadsheet its power.

Although the modern spreadsheet program is highly sophisticated with advanced facilities for statistics and programming, it is possible to benefit enormously from their use while hardly scratching the surface of their functionality. In particular, spreadsheets can make collecting and processing results for experiments very convenient. They completely avoid the need to ever start over because of some stupid mistake in a data value or a formula. Just edit the mistake and everything dependent on it is recalculated.

When it comes to plotting a graph, a spreadsheet program like Microsoft Excel makes it simplicity itself, as it is generally just a question of selecting the data you want plotted and then the type of graph you would like. As with derived values, changing a data value will immediately update any graphs that are dependent on the changed values.

## Example Spreadsheet

The following will use the Charles's law example to show how Excel can be used. Start by giving your sheet a heading and then putting any constants onto the sheet, as shown in Figure A2.1.

You enter values by clicking on the cell and then typing in the value. Here we have the one constant, the radius of the tube. Notice that I have put a label for the value to the left, and the units to the right of the cell that contains the actual value.

Enter the values that you find while doing the experiment, as shown in Figure A2.2.

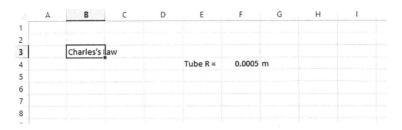

**FIGURE A2.1**
An Excel spreadsheet with some initial values.

| | A | B | C | D | E | F | G |
|---|---|---|---|---|---|---|---|
| 1 | | | | | | | |
| 2 | | | | | | | |
| 3 | | Charles's Law | | | | | |
| 4 | | | | | Tube R = | 0.0005 m | |
| 5 | | | | | | | |
| 6 | | T° C | L cm | | | | |
| 7 | | | | | | | |
| 8 | | 95 | 7.28 | | | | |
| 9 | | 90 | 7.2 | | | | |
| 10 | | 85 | 7.08 | | | | |
| 11 | | 80 | 6.98 | | | | |
| 12 | | 75 | 6.87 | | | | |
| 13 | | 70 | 6.78 | | | | |
| 14 | | 65 | 6.67 | | | | |
| 15 | | 60 | 6.57 | | | | |
| 16 | | 55 | 6.47 | | | | |
| 17 | | 50 | 6.38 | | | | |
| 18 | | 45 | 6.27 | | | | |
| 19 | | 40 | 6.18 | | | | |
| 20 | | 35 | 6.08 | | | | |
| 21 | | 30 | 5.97 | | | | |
| 22 | | 25 | 5.88 | | | | |
| 23 | | | | | | | |
| 24 | | | | | | | |

**FIGURE A2.2**
An Excel spreadsheet with the experiment's observed values.

Create a column for the volume of the gas. Now enter the formula for the volume into the first row for the volume. You start a formula for a particular cell by clicking on the cell (here it is D8) and then press equals (this tells Excel that what is coming is a formula rather than a value). You then type in the formula. In this case it is:

$$=PI()*\$F\$4*\$F\$4*C8/100$$

Which might, at first sight, seem a little complicated. However, it is really quite simple. The volume of a cylinder is given by:

$$V = \pi\, r^2\, L$$

The formula starts with Excel's way of getting the value for pi, which for Excel is a mathematical function called PI. It has opening and closing brackets because functions sometimes have arguments or parameters that will usually be given between opening and closing brackets. Here there are no parameters, but we still have the brackets.

This is followed by a multiplication with the radius twice; this is the r-squared part (it is actually faster for the computer to work out a square by using multiplication than to invoke exponentiation using the '^' symbol).

The dollar signs are used because when we copy and paste this formula to calculate all the other rows, we want this cell address to be treated as an absolute address and not changed in any way.

Finally, we have the multiplication by C8, which is the length in centimeters, so we divide by 100 to have our result in meters cubed.

Now for the interesting bit. Note that in Figure A2.3, the D8 cell is selected, and it has a black box in the lower right-hand corner. This can be dragged

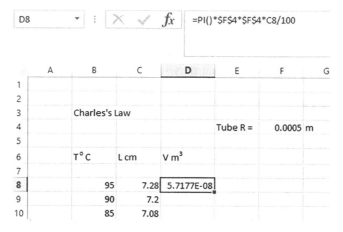

**FIGURE A2.3**
An Excel spreadsheet showing an entered formula.

| | T°C | L cm | V m³ |
|---|---|---|---|
| 6 | | | |
| 7 | | | |
| 8 | 95 | 7.28 | 5.7177E-08 |
| 9 | 90 | 7.2 | 5.6549E-08 |
| 10 | 85 | 7.08 | 5.5606E-08 |
| 11 | 80 | 6.98 | 5.4821E-08 |
| 12 | 75 | 6.87 | 5.3957E-08 |
| 13 | 70 | 6.78 | 5.325E-08 |
| 14 | 65 | 6.67 | 5.2386E-08 |
| 15 | 60 | 6.57 | 5.1601E-08 |
| 16 | 55 | 6.47 | 5.0815E-08 |
| 17 | 50 | 6.38 | 5.0108E-08 |
| 18 | 45 | 6.27 | 4.9244E-08 |
| 19 | 40 | 6.18 | 4.8538E-08 |
| 20 | 35 | 6.08 | 4.7752E-08 |
| 21 | 30 | 5.97 | 4.6888E-08 |
| 22 | 25 | 5.88 | 4.6181E-08 |
| 23 | | | |
| 24 | | | |

**FIGURE A2.4**
An Excel spreadsheet showing the computed results.

downwards, and the correct formulae will be copied into all the cells that you drag to. You should get what is shown in Figure A2.4.

Now you can appreciate the need for absolute addressing used in the formula. Each use of the cell F4 is absolute; it is always in F4 and nowhere else. If we had not made it absolute, then the pasted formulae would have created offsets from F4.

You may be wondering why these values are so very small. It is because the units of meters cubed are very large when we are calculating volumes in a capillary tube. However, it is always worth doing a quick visual check as we enter formulae to ensure that they make sense.

## Creating the Graph

Select all three columns of data by dragging from the top left of the data to the bottom right. With the Insert tab selected from the top of the Excel interface, select the scatter plot icon as shown in Figure A2.5.

The small window of scatter chart options that appears is shown in Figure A2.6.

Choose the first of the available scatter plots and you will get this graph placed on your sheet, as shown in Figure A2.7.

**FIGURE A2.5**
Selecting the scatter plot.

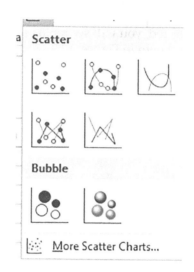

**FIGURE A2.6**
Selecting the correct scatter plot.

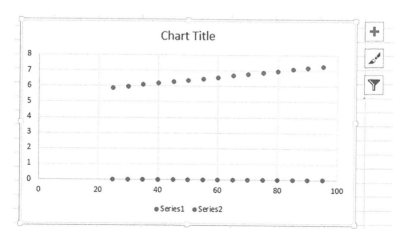

**FIGURE A2.7**
The scatter plot on the spreadsheet.

This is not exactly what we want, so we need to do a few things. Firstly, we seem to have two plots. This is because we chose all three columns of data. From the data and the plots, it is clearly the top one that we do not want. Click on one of the blue data points on the graph and you will see this confirmed on the spreadsheet by those columns of data becoming highlighted.

Press the delete key to get rid of this plot. Now you should be looking at what is shown in Figure A2.8.

This confirms that you have plotted the correct values. Get rid of the series two label by clicking on it and pressing delete. We do not need it, as we only have one series. We can now change the chart title by clicking on it and entering some new text.

When the chart is selected, you will see a plus sign at the top right. Click on this to see this menu shown in Figure A2.9.

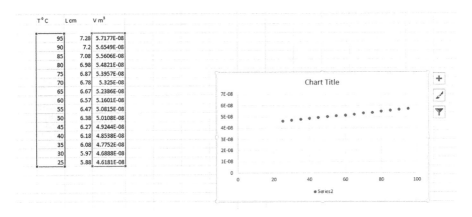

**FIGURE A2.8**
The amended scatter plot on the spreadsheet.

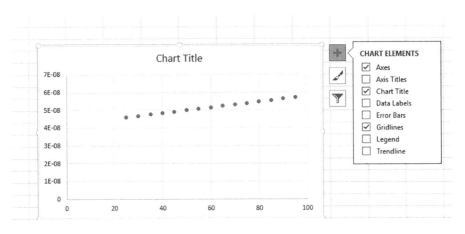

**FIGURE A2.9**
The Chart Elements menu.

Click on the Axis Titles checkbox and then edit them to give you appropriate labels for both axes. Now we want to add a best-fit line. From the menu, click on Trendline and on the small right arrow that appears, click on More Options. This will cause a panel to be displayed on the right with all the Trendline options displayed, as shown in Figure A2.10.

Select the linear option, as we want a straight line. It would also be rather nice to have the equation of the line displayed, so click on that option (near the bottom). You may need to drag the label away from the line in order to be able to read it.

Note that the equation, as displayed, has too few decimal places to be of any use. Right-click on the equation label and choose Format Trendline Label from the displayed menu. Under Category, choose Scientific, and specify decimal places as 2. Now we have our graph with a usable equation for the best-fit straight line, as shown in Figure A2.11.

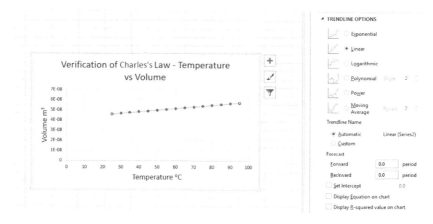

**FIGURE A2.10**
The trendline options.

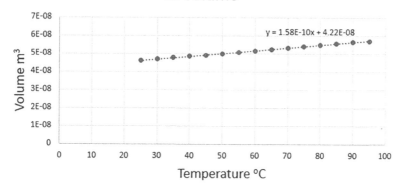

**FIGURE A2.11**
The graph with the labeled trendline.

# Appendix 3

## Controlling the Simulations

The important controls are:

- Move yourself using cursor keys and PageUp/Down to change height
- Change view/direction by dragging mouse with right button held down
- Click on switches to change state
- Use mouse wheel or P and M when mouse cursor is over objects, on rotating knobs, and to move items
- Use +/– or Z and X to zoom in and out
- Where there is a slide show, use switches on projector to change slide being projected

In addition, some objects can be picked up and moved using a single mouse click followed by movement of the mouse. When this is possible, movement towards and away from you can be affected by using shift with the mouse wheel. You can let go of an object by using a second mouse click, or by colliding the object against anything else.

All the experiments also have the option of using on-screen controls. These are displayed by clicking on the button at the bottom center of the screen; see Figure A3.1.

The left-hand joystick controls movement, while the right-hand joystick controls where you are looking and simultaneously defines the direction of straight ahead. The magnifying glass icons allow you to zoom in and out, and the eye-up and -down icons allow you to move upwards and downwards.

If you are trying to move an object, click on it to drag it, then move yourself using the movement joystick and the object will move with you.

In the on-screen controls mode, you can rotate knobs by dragging on them.

**FIGURE A3.1**
The on-screen controls.

# Back Cover

*High School and Undergraduate Physics Practicals* describes more than thirty physics practicals at high school and undergraduate levels with background information on each one, a description of the equipment needed, and instructions on how the experiment is performed. Uniquely, for those without access to a real laboratory, the book gives you access to highly detailed 3D simulations of all the experiments.

The simulations are a superset of the Virtual Physics Laboratory as reviewed and given the Green Tick of Approval by the Association for Science Education. They run in any browser that supports WebGL, such as Microsoft Edge or Firefox on Windows and Safari on the Mac. For the school or university student who wants to practice and widen their knowledge of physics, or for those who are learning on their own, this is an ideal book for honing and broadening experimental skills.

The simulations are the result of many years researching the teaching of online science, a field in which the author has published many papers.

# Index